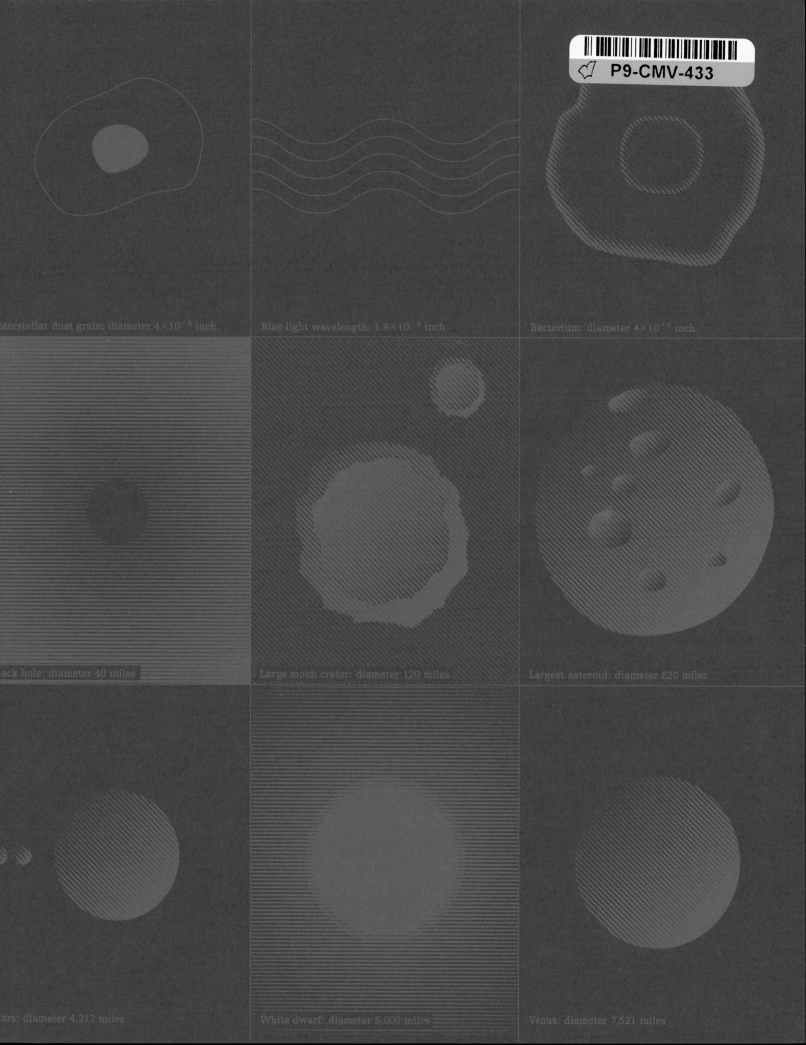

Interstellar dust grain: diameter 4×10^{-5} inch

Blue light wavelength: 1.9×10^{-5} inch

Bacterium: diameter 4×10^{-5} inch

Black hole: diameter 40 miles

Large moon crater: diameter 120 miles

Largest asteroid: diameter 620 miles

Mars: diameter 4,217 miles

White dwarf: diameter 5,000 miles

Venus: diameter 7,521 miles

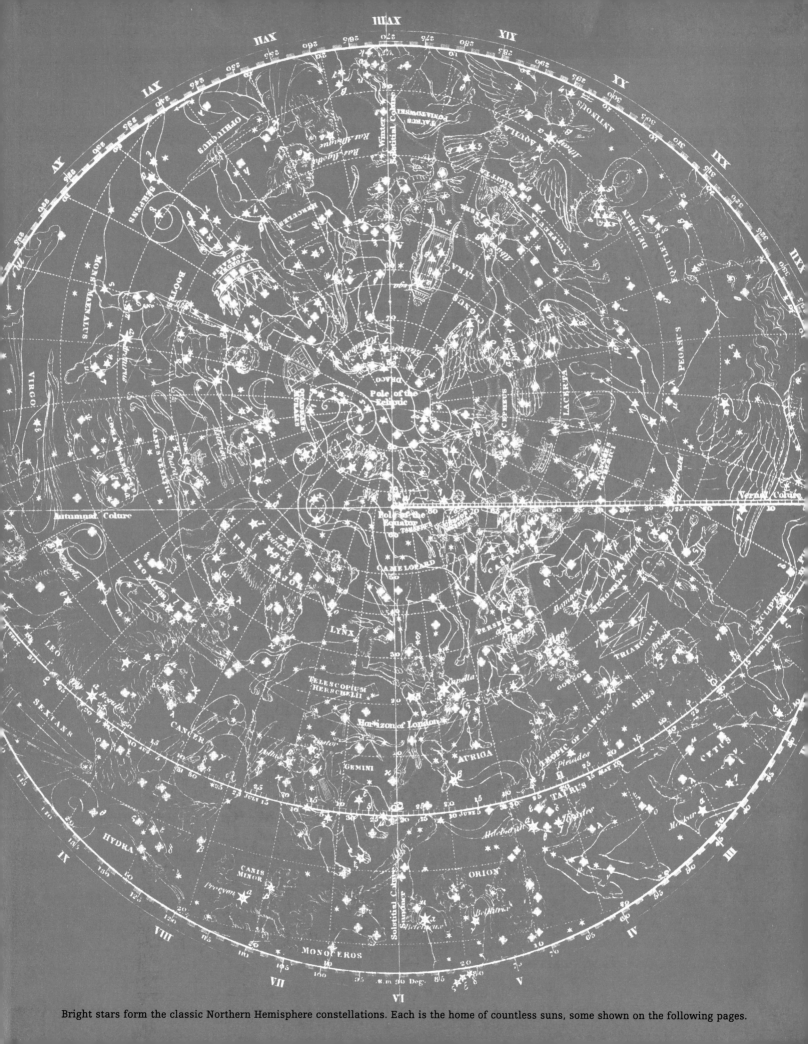

Bright stars form the classic Northern Hemisphere constellations. Each is the home of countless suns, some shown on the following pages.

Zeta Orionis *(below)*, leftmost star in Orion's belt *(red inset)* and 35,000 times brighter than the Sun, blazes beside a nebulous complex that includes a thick coil of dust and gas known as the Horsehead *(below, right),* a nursery where prestellar bodies called protostars may now be coalescing.

On the shoulder of Taurus the Bull perch the Pleiades, young stars glowing through the lingering dust of their formation.

Nestled in a blue cloud spawned by its own expulsion of matter, a rare Wolf-Rayet star nicks the ear of Canis Major.

Near the jaw of the Unicorn, the Rosette nebula's pink ionized hydrogen enfolds a clutch of blue stellar infants 500,000 years old.

Its yellow-red light diffused through dust, Antares *(left)*, fifteenth-brightest star in the sky, helps form the southern constellation Scorpius.

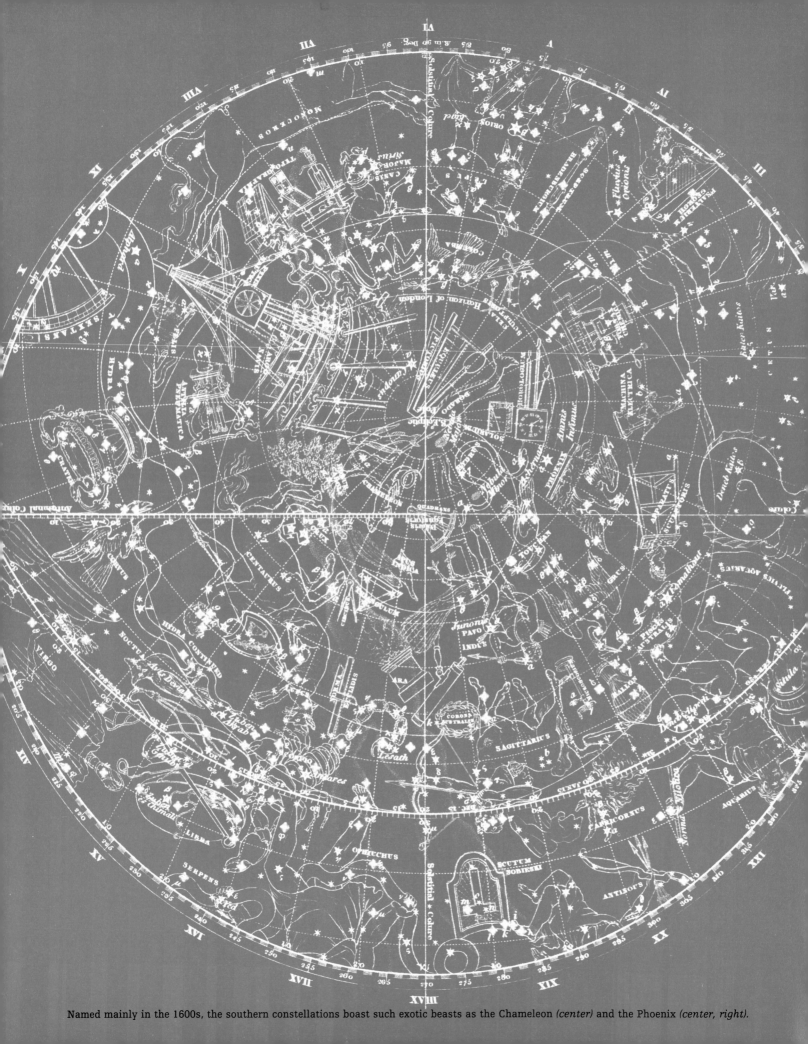

Named mainly in the 1600s, the southern constellations boast such exotic beasts as the Chameleon *(center)* and the Phoenix *(center, right)*.

STARS

Other Publications:
AMERICAN COUNTRY
THE THIRD REICH
THE TIME-LIFE GARDENER'S GUIDE
MYSTERIES OF THE UNKNOWN
TIME FRAME
FIX IT YOURSELF
FITNESS, HEALTH & NUTRITION
SUCCESSFUL PARENTING
HEALTHY HOME COOKING
UNDERSTANDING COMPUTERS
LIBRARY OF NATIONS
THE ENCHANTED WORLD
THE KODAK LIBRARY OF CREATIVE PHOTOGRAPHY
GREAT MEALS IN MINUTES
THE CIVIL WAR
PLANET EARTH
COLLECTOR'S LIBRARY OF THE CIVIL WAR
THE EPIC OF FLIGHT
THE GOOD COOK
WORLD WAR II
HOME REPAIR AND IMPROVEMENT
THE OLD WEST

This volume is one of a series that
examines the universe in all its aspects,
from its beginnings in the Big Bang to the
promise of space exploration.

VOYAGE THROUGH THE UNIVERSE

STARS

BY THE EDITORS OF TIME-LIFE BOOKS
ALEXANDRIA, VIRGINIA

CONTENTS

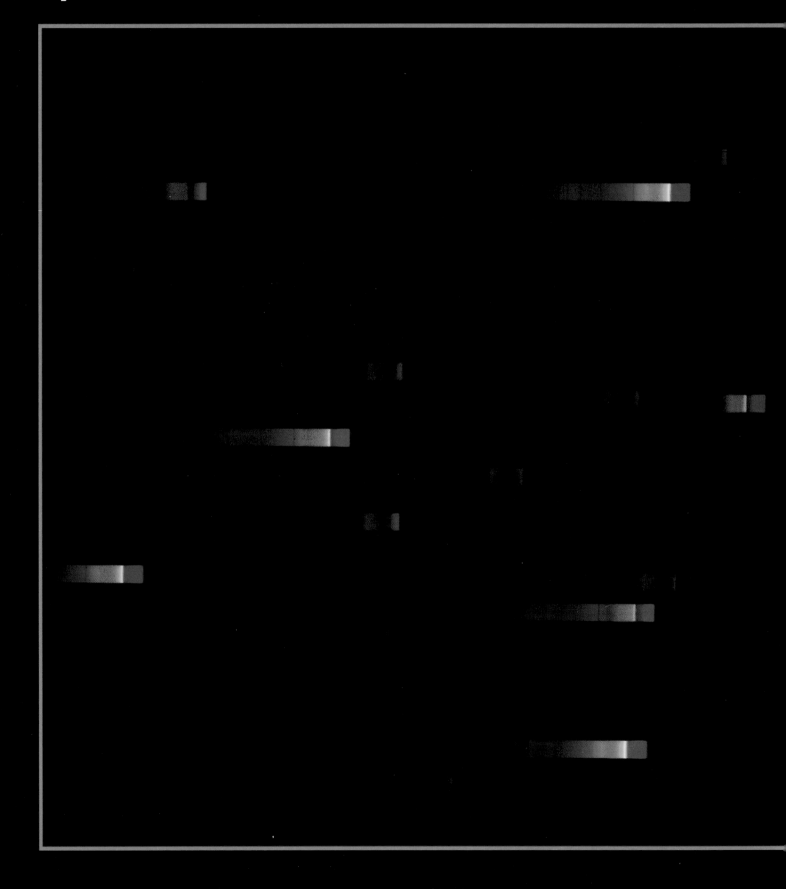

Photographed through a telescope equipped with a prism, the Hyades star cluster is transformed into a collage of stellar rainbows that are known as spectra, each crossed by fine dark lines that hint at the stars' inner secrets.

When day ends, the blazing Sun relinquishes the sky to the pinprick glimmerings of a multitude of cousins. The appearance of these other suns triggers a flurry of activity on the nightside of the globe as astronomers perform last-minute calibrations—pushing buttons, flicking switches, and sighting through the eyepieces of their telescopes. Then, guided by computers, the giant contraptions of steel and glass begin tracking the stars. But once the telescope locks onto its target, human eyes seldom bother to look again. Even through the most powerful instrument, stars are too far away—the nearest being some 26 trillion miles distant—to show up as more than specks of light. What astronomers covet is not a visual glimpse of these celestial bodies but a more permanent record of their elusive radiation. The nineteenth-century scientists who broke the code of starlight—who learned that this tenuous glow carries information about a star's basic nature—opened up the heavens as surely as Columbus opened up the Western world more than three centuries earlier.

Subsequent generations have discovered that stars are numerous beyond comprehension. From Earth, the naked eye can glimpse as many as 6,000 of them. But the Milky Way alone probably contains 200 billion of these great spheres of shining gas, and it is only one galaxy among at least 10 billion in a universe so vast that light—traveling at 186,000 miles per second—would take some 30 billion years to cross from one side to the other. The English astronomer Sir James Jeans may not have been exaggerating when he asserted that the universe has as many stars as there are grains of sand on all the ocean beaches of the Earth.

Scattered like diamond chips across the cosmos, the stars look deceptively serene to the earthbound observer. Yet each one is a creature of extraordinary violence. Temperatures on the surfaces of normal stars can range up to 50,000 degrees Kelvin, more than ten times that of the Sun; in some rare stars, the surface seethes at an unimaginable one million degrees. (The Kelvin scale, always used for stellar temperatures, corresponds to Celsius but sets zero degrees at absolute zero, or minus 273 degrees on the Celsius scale.) Stellar interiors burn with even greater intensity: They are nature's own thermonuclear furnaces, raging at several million degrees Kelvin in typical stars—

so hot that atomic nuclei are torn apart and made into new types of matter.

The peaceful-looking heavens disguise not only the ferocity but the enormous variety of stars as well. Often migrating through the galaxy in binary pairs, triplets, and even quartets *(page 18)*, stars range in size from smaller than the Earth to nearly as large as the Solar System. Some regularly expand and contract, growing brighter and dimmer as they do so. For example, Betelgeuse, the brightest star in the constellation Orion, varies in diameter by as much as 170 million miles over a six-year period. Although their lives may be measured in billions of years, stars do age and die. Some go quietly, cooling into dark remnants of their former splendor. Others collapse and explode in a shattering cataclysm. But even as old stars vanish, new ones are being born in the gaseous stellar nurseries of the universe.

STRIPED LIGHT

Almost everything astronomers now know about the bewildering variety of stars is based on a technique known as spectroscopy *(pages 33-39)*, the analysis of light and other radiation. When the information in starlight became available to astronomers, an unprecedented era of knowledge gathering began—a time when scientists analyzed and recorded and classified hundreds of thousands of stars, laying the groundwork for twentieth-century breakthroughs in stellar theory.

Like much of modern astronomy, spectroscopy owes its beginnings to the penetrating imagination of Isaac Newton, the extraordinary seventeenth-century English mathematician and scientist. Newton subscribed to the strange notion, proposed by earlier thinkers such as René Descartes, that white light holds all the colors of the rainbow. In 1666 he performed his first experiment—using a glass prism, a little round hole in one of his window shutters, and the white wall of the room—"to try therewith," as he wrote later, "the celebrated Phaenomena of colours." He permitted a stream of sunlight to enter the darkened room through the hole and pass through the prism. The glass dispersed the light into an array of slightly overlapping colors: from red, through orange, yellow, green, and blue, to violet. To describe the blurry, multihued image that appeared as if by magic on his wall, Newton borrowed from Latin the word *spectrum,* meaning "apparition" or "specter."

Although the experiments brought Newton at age twenty-three his first taste of fame, the results were not immediately incorporated into the infant science of astronomy. As late as 1790, the German cultural luminary Johann Wolfgang von Goethe, who aspired to be a scientist as well as a poet and philosopher, wrote, "The idea of white light being composed of colored lights is quite inconceivable, mere twaddle, admirable for children in a go-cart." Astronomers in any case showed little interest in the composition of stars or starlight during this period. Peering through their telescopes, they mapped stellar positions and attempted to chart stellar motions. But eighteenth- and nineteenth-century scientists thought of the stars as a mere backdrop for what really fascinated them: the dance of the planets at center stage.

Part of this disregard may have stemmed from the widespread skepticism that science could ever decipher the true physical nature of the stars—a skepticism that would largely evaporate thanks to discoveries made by a gifted German optician named Joseph von Fraunhofer. How Fraunhofer became one of Germany's most noted nineteenth-century scientists is a tale worthy of Dickens. The eleventh child of an impoverished master glazier, he was a lively and inquisitive boy who was orphaned before the age of twelve and then apprenticed to an unimaginative mirror maker and glass cutter in Munich. It took a near calamity to save the youth from a life of misery. When Fraunhofer was fourteen, the slum building in which he lived and worked suddenly collapsed, pinning the young apprentice and killing everyone else. After he was pulled from the wreckage, a Bavarian prince, Maximilian Joseph, heard of his ordeal and gave him eighteen gold ducats. It was enough to enable Fraunhofer to purchase a glass-working machine, books on optics, and early release from his hated apprenticeship.

In 1806 Fraunhofer joined a Munich firm known for the quality of its optical instruments. Within five years he had become so skilled at glass making, lens grinding, and design that, at age twenty-four, he was made a full partner in the firm. His pursuit of the ideal lenses for telescopes and other instruments led the young perfectionist to experiment with spectroscopy. In 1814 he set up a telescope taken from one of his surveying instruments and let sunlight enter the room through a narrow slit in the window shutter.

Through the eyepiece, he observed the spectrum created when the light was diffracted by a prism mounted in front of the telescope. He wrote later that the rainbow of colors appeared as expected. But then he saw something else: "an almost countless number of strong and weak vertical lines, which are, however, darker than the rest of the color image; some appeared to be almost perfectly black." These fine dark stripes would later become familiar to every student of physics as the Fraunhofer absorption lines. Newton had not seen them, probably because the round hole he used to concentrate his light beam was, unlike Fraunhofer's narrow slit, too large to produce clearly defined dark spaces in the spectrum. Another of Fraunhofer's predecessors, the English chemist William Wollaston, had observed the lines in 1802 but erroneously concluded that they were merely gaps separating the bands of color.

Perplexed by the origin of the lines but certain they were inherent in solar light and not artifacts of his experiments, Fraunhofer studied them assiduously. He mapped some 600 in the solar spectrum (more than 20,000 are known today) and gave the most prominent ones alphabetical labels, some of which survive in modern astronomy. He then turned his new instrument, the spectroscope, on the Moon, Venus, and Mars. The patterns of lines in all three spectra appeared identical to those in the solar spectrum, and Fraunhofer rightly concluded that this was because these bodies were not inherently luminous; instead they reflected the light of the Sun. When he focused on Sirius and five other bright stars, however, he found that starlight told a different story. Each star's spectral pattern was different, and all differed

Unlike the solitary Sun, most stars in the Milky Way galaxy belong to binary or multiple systems. Because the binary orbit is more stable than others, multiple systems are often made up of pairs, such as the visible members of Beta Cygni *(near left)*. The Sigma Orionis trio *(far left, top)* includes a binary, a third star orbiting those two at a distance, and perhaps two hidden companions. Theta Orionis's four bright stars *(far left, bottom)*—and several members too dim to be seen here—are relatively equidistant, and astronomers consider their tangled orbits highly unstable; the system may fly apart in just a few million more years.

from the pattern for sunlight. Apparently these patterns were a kind of stellar fingerprint that identified each star and set it apart from all others.

In the process of studying the dark lines, Fraunhofer vastly improved a device known as a diffraction grating, an alternative to the prism for displaying spectra. With the aid of a machine he invented, Fraunhofer developed a grating that consisted of 3,200 lines etched with a diamond point into a piece of glass less than half an inch wide. Such gratings yielded far more detailed spectra than was possible with the prism, enabling the scientist to measure with some precision the relative positions of his mysterious dark lines.

Fraunhofer tested his simple spectroscopes—a term coined later to describe devices for viewing the spectrum—by passing both sunlight and light from gas flames through them. In doing so, he identified a variety of spectral line that appeared only when he studied light from laboratory flames. These lines, which were bright and colored instead of dark, came to be known as emission lines because they are emitted by substances heated to incandescence. Despite this difference, Fraunhofer noted a coincidence between the positions of a pair of dark lines in the solar spectrum and a pair of bright yellow ones in the spectra of chemicals burned in his lab.

Fascinating as these experiments were, the master telescope builder was so devoted to his craft that he spent little time on the riddle of the lines. Before he could solve it, he contracted tuberculosis and died in 1826, ending at thirty-nine a life of remarkable scientific achievement. In addition to helping to establish the German optical industry as a world leader, Fraunhofer left behind important clues to the nature and function of the dark spectral lines. For one thing, he speculated that they were caused by the absence of particular wavelengths of light, as if something in the Sun and other stars had robbed their spectra of narrow stripes of color.

BREAKING THE CODE

Researchers who came after Fraunhofer tended to focus on the bright rather than the dark spectral lines, and on the laboratory rather than the heavens. Although this work proceeded in a number of countries over the next three decades, the key experiments were performed in 1859 at the University of Heidelberg by a pair of Fraunhofer's countrymen, Gustav Kirchhoff and Robert Bunsen. Unlike the self-made Fraunhofer, both scientists were the offspring of middle-class privilege. Kirchhoff was a trained physicist who had grown up in a thriving intellectual atmosphere in Königsberg as the son of a government law counselor. He had earned a considerable reputation in the field of electrical theory by demonstrating that electrical impulses move at the same speed as light. Bunsen, Kirchhoff's senior by thirteen years, was a native of another German intellectual center, Göttingen, and the inventor of such instruments as the Bunsen burner. A laboratory explosion that had cost him his sight in one eye and a near-fatal experiment with arsenic had not deterred him from a career of careful chemical research.

The pair's collaboration grew out of Bunsen's attempts to identify chemicals by the colors with which they burned. Kirchhoff suggested to his friend and colleague that he could make the clearest distinctions by passing light from Bunsen's new burner through a spectroscope. The device was especially useful for these experiments because it produced a hot flame that was free of impurities and yielded spectra with sharply defined lines. Soon it became apparent that each chemical element, when burned as a gas, created a unique pattern of spectral emission lines. Sodium, for example, produced the double yellow lines observed by Fraunhofer, who had seen it in so many spectra because traces of the element are found in numerous substances.

It was a momentous breakthrough. The discovery that each chemical element carried its own spectral signature had enormous implications for chemistry, physics, and astronomy. Through spectral analysis, Bunsen and Kirchhoff identified the characteristic patterns of colored emission lines for all the elements then known.

Kirchhoff, meanwhile, began conducting experiments that formed the bedrock of a branch of astronomy later known as astrophysics. First, he solved the old mystery of the origin of the lines in solar and stellar spectra. Confirming the coincidence noticed by Fraunhofer, he established that certain solar absorption lines, now known as Fraunhofer's D lines, matched the emission lines of sodium. He trained the spectroscope on a flame of sodium burning against a dark background to produce the double yellow emission lines characteristic of that element, then moved the same flame into the path of a beam of sunlight, thinking that the bright emission lines would cancel out the matching absorption lines in the Sun's spectrum. Instead, the absorption lines looked clearer and darker than before. Apparently, the gas was absorbing much more energy from the sunlight than it was emitting.

Kirchhoff applied these principles to astronomy with stunning effect. Light from the hot Sun or from another star passes through a surrounding atmosphere of cooler gases, Kirchhoff concluded. Gases such as sodium vapor absorb their characteristic wavelengths from the light, producing the dark Fraunhofer lines in the spectrum reaching the Earth. To demonstrate, Kirchhoff rigged a telescope with a four-prism spectroscope that let him simultaneously view and compare the dark lines in the solar spectrum with the bright emission lines for thirty different elements found on Earth. Deciphering the composition of the solar atmosphere, he found not only sodium but also significant amounts of iron, calcium, magnesium, nickel, and chromium. A few years later, astronomers viewing the Sun during a solar eclipse even found the spectral lines of an element not yet discovered on Earth. They named it helium, after the Greek word *helios*, for "sun."

FOUR TYPES OF STARS
Equipped with the new tool of spectral analysis, astronomers began to mine nuggets of information from hundreds of stars other than the Sun. Among the eager researchers was a wealthy Englishman named William Huggins. Working in the observatory he had built atop his London home, Huggins showed

that the stars, like the Sun, contained elements such as sodium, iron, and calcium that are commonly found on Earth. Thus, in 1863 he could announce with some confidence that the Earth and heavens were one: "The stars, while differing the one from the other in the kinds of matter of which they consist, are all constructed upon the same plan as our sun, and are composed of matter identical at least in part with the materials of our system."

As Huggins and others decoded more stellar spectra, other researchers developed increasingly detailed systems for classifying the patterns. Astronomers saw that stars fit into categories based on their brightness and color and the characteristic imprint of their spectra. These categories would in turn give the scientists a larger picture of stellar life, death, and kinship.

The first successful system of classification emerged in 1868 from an unlikely source: the Vatican. Known earlier for its persecution of astronomers, the papacy now sponsored a first-class research center in Rome, the Roman College Observatory. Its director was Father Angelo Secchi, a fifty-year-old Jesuit and physicist. Secchi had learned the astronomer's trade in England and the United States in the 1840s, when his religious order was temporarily exiled by the Italian government. Working in the unfinished dome of the church of San Ignazio on the grounds of the observatory, Secchi set out in 1863, as he wrote later, "to see if the composition of the stars is as varied as the stars are innumerable." He examined the spectra of some 4,000 stars over a five-year period and found that the answer was yes . . . but.

Yes, stars were breathtakingly diverse, but Secchi saw enough similarities to classify them into four main spectral types. These categories were linked

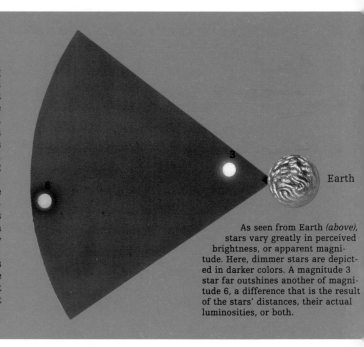

A QUESTION OF BRIGHTNESS

As early as 120 BC, Greek astronomers ranked the stars according to six categories of brightness, from dazzling objects they named "first magnitude" to stars of the "sixth magnitude" that were just visible to the naked eye. Modern observers employ much the same system but have extended the categories to include extremely faint, twenty-seventh magnitude objects that can be seen only through sophisticated telescopes, as well as very brilliant bodies like the Sun or full Moon, whose magnitudes are measured in negative numbers. Each order of magnitude is strictly quantified to be exactly 2.512 times as dim as the one before.

As people learned more about the stars and starlight, it became clear that such magnitudes did not reflect a star's intrinsic brightness but rather combined the star's actual light output with its distance from Earth. Placed far enough away, even the blazing Sun would become a dim point of light; brought near, a once-faint body might appear a million times as bright.

Thus, a star's brightness as seen from Earth became known as its "apparent" magnitude, and astronomers devised a new measure called "absolute" magnitude, calculated by determining how bright a star would appear at a standard distance from Earth—the truest measure of stellar brilliance.

Earth

As seen from Earth (above), stars vary greatly in perceived brightness, or apparent magnitude. Here, dimmer stars are depicted in darker colors. A magnitude 3 star far outshines another of magnitude 6, a difference that is the result of the stars' distances, their actual luminosities, or both.

by star color as seen through the telescope and by the position, width, number, and strength, or darkness, of absorption lines in the spectrum. Type I stars were white or blue, like Sirius, with strong hydrogen absorption lines. Type II included the Sun—yellow or orange stars with numerous lines indicating the presence of metallic elements such as iron. Type III were orange to red, with wide bands made up of many fine lines grouped together. Type IV, the deepest red stars, had carbon absorption bands in their spectra; few of these stars were visible to the naked eye.

Fortunately for astronomers who followed Secchi, the camera soon overtook naked-eye observation in spectral analysis. The sensitivity of the photographic plate to ultraviolet light permitted researchers to obtain images of starlight too faint to be seen through the telescope and to record wavelengths invisible to human sight. And instead of sitting up all night peering through an eyepiece to catch the faint glimmer of a spectrum, scientists could study and measure spectral portraits on photographic plates at their leisure.

The first researcher to successfully photograph a stellar spectrum was Henry Draper, a prosperous New York physician, academician, and amateur astronomer. Draper established his own observatory at Hastings-on-Hudson, New York, on the estate of his father, John William Draper. The elder Draper was himself a physician, astronomer, and photographer who in 1843 had taken one of the first photographs of the solar spectrum. Henry photographed the spectrum of the star Vega in 1872 and went on to record the spectra of more than eighty other stars with a device he called a spectrograph—a spectroscope equipped with a camera. His research was cut short in 1882 when he died of double pleurisy at age forty-five. Four years later, his widow, Mary Anna Draper, established the Henry Draper Memorial Fund to help finance an ambitious program of stellar spectrography and classification at the Harvard College Observatory.

In order to compare the true luminosity of stars, astronomers calculate their absolute magnitudes—how brightly each would shine at a uniform distance of ten parsecs, an astronomical measure equivalent to about 32.6 light-years *(white curve)*.

Earth

Moved mathematically to the standard distance, the nearby star drops from an apparent magnitude of 3 to an absolute magnitude of 6. The distant star, pulled in to the line, jumps to an absolute magnitude of 5, a value that marks it as intrinsically more luminous than the other star.

NAMING THE STELLAR FAMILY

The Harvard project was the brainchild of a remarkable man who served as the observatory's director for forty-two years, Edward C. Pickering. After spending a decade as a professor of physics at the Massachusetts Institute of Technology, Pickering crossed town to Harvard in 1877. Already known as an innovative teacher, this descendant of one of New England's oldest and most distinguished families proved to be an energetic administrator as well, a tireless fund-raiser willing to dip into his own pockets when necessary. He was also a "collector of astronomical facts," as he styled himself.

Soon after Mrs. Draper's first gift to the fund—in all she contributed nearly $400,000—Pickering set out to collect as many astronomical facts as possible. He rigged the observatory's wide-angle telescopes with special prisms that spread the spectra of hundreds of stars simultaneously across a single eight-by-ten-inch photographic plate. And to bring order to the plates that piled up in the observatory's brick headquarters on the outskirts of Cambridge, he hired a corps of women employees. Other astronomers joked about "Pickering's harem," but the director's motives were purely pragmatic. He believed women were ideally suited by temperament to the tedious calculations and measurements necessary to determine the wavelengths of the spectra on the black-and-white plates. Whatever the truth of that prejudice, he also knew women would work for considerably less money than men.

Pickering's chief assistant in charge of managing the staff was Williamina Fleming, a native of Scotland and a strict disciplinarian who generated respect bordering on awe among her subordinates. Fleming had worked as Pickering's housekeeper until 1881, when the director, irritated by the sloth of a male assistant, vowed that his Scots maid could do better and promptly gave her the job. Fleming was so devoted to Pickering that she named her son after him, but even she chafed privately under her boss's double standard. "He seems to think that no work is too much or too hard for me, no matter what the responsibility or how long the hours," she complained in her diary. "But let me raise the question of salary and I am immediately told that I receive an excellent salary as women's salaries stand."

In 1886, before she was promoted to chief assistant, Fleming undertook the observatory's first classification project. Developed jointly with Pickering, the project enlarged and refined Father Secchi's old system based on similarities in spectral line patterns. Fleming and Pickering subdivided Secchi's original four types into thirteen classes denoted by alphabetical letters (skipping over J because it was too easily confused with I). For example, stars in the first class, A, were characterized by spectra with strong hydrogen lines; in types B, C, and D, the hydrogen lines grew progressively more faint. Pickering also added three new classes, labeled O, P, and Q, to cover newly recognized and miscellaneous spectra. Fleming's work resulted in the most extensive classification effort to date, a catalog of 10,351 stars of the Northern Hemisphere, published in 1890.

As spectroscopic methods improved, Pickering assigned a second project; he wanted more detailed and in-depth studies of 681 bright northern stars. This job went to Antonia Maury not long after she joined the staff as a research associate in 1888. The assignment seemed appropriate, for Maury was the niece of Henry Draper, whose memorial fund helped finance the work and whose old eleven-inch telescope was used to record the 4,800 plates of stellar spectra she would study. Moreover, Maury had been well trained in physics and astronomy. At Vassar College, she had come under the influence of Maria

Vine-covered buildings crowd the grounds of the Harvard College Observatory in this 1897 photograph. Observatory staff relied on several refracting telescopes, including the eight-inch Draper refractor, housed in the open-roofed structure at far left; the thirteen-inch Boyden instrument, in the full dome to its right; the eleven-inch Draper telescope, glimpsed within the dome at center; and at top right, an unnamed fifteen-inch refractor.

Mitchell, America's first woman astronomer, and had graduated with honors. Almost from the beginning, however, she and Pickering came into conflict. A strong individualist, Maury refused to blend into his professional harem. In an era when women were expected to be conscious of conservative fashion, she paid little attention to her appearance and at times showed up at the office in mismatched stockings. Not infrequently she bore the marks of insect bites incurred because she insisted on working at night with the windows open and could not bear to kill the mosquitoes. If such idiosyncrasies were not enough for Pickering, there was the influence of Maury's elegantly turned-out aunt, Mary Anna Draper, who made plain to him her own feelings of distaste for her dowdy relative by marriage.

The main source of friction, however, was what one of Maury's colleagues described as her "passion for understanding." Pickering, aiming for mass production of the stellar spectra, wanted quick results; Maury was not content merely to classify but, as the colleague said, "was always slowing things up by asking what it meant." For her classification project, she abandoned the system used by Williamina Fleming and developed a far more complex scheme. She created twenty-two groups, designated by Roman numerals, that generally paralleled Pickering's sixteen alphabetical classes. Within each group, moreover, she defined subdivisions, using the letters a, b, and c, to distinguish the relative strengths and physical appearances—whether wide or narrow, hazy or sharp—of certain spectral lines.

Pickering was unsympathetic to such subtle distinctions. Fed up with his

unrelenting emphasis on quantity, Maury left Harvard in 1896, a year before the publication of her spectra study, and stayed away for more than two decades, teaching in private schools. However, several years after her departure from Harvard, other astronomers realized that Maury had made an important contribution by isolating, without knowing it, certain kinds of stars. One spectral type, for instance, was later found to signal rapidly rotating stars; another indicated huge stars, later known as supergiants.

Meanwhile, Pickering had found a protégé far more to his taste. Like Maury, the new woman, Annie Jump Cannon, arrived well versed in astronomy and physics. As a child in Delaware, where her father served in the state senate, she had had her own makeshift observatory in the attic of their comfortable home, and she later studied at both Wellesley and Radcliffe. In Pickering's opinion, Cannon possessed much that Maury lacked: She was gaily dressed and of a cheerful disposition, laughing often and heartily. Even the hearing aid she wore proved to be a virtue; if she wanted to concentrate undisturbed on her work, she simply unplugged it. Best of all, as far as her boss was concerned, she was happy to classify the spectra without getting bogged down, Maury style, in matters of interpretation. "Miss Cannon was not given to theorizing," wrote Cecilia Payne-Gaposchkin, her colleague at the observatory in later years. "It is probable that she never published a controversial word or a speculative thought. That was the strength of her scientific work— her classification was dispassionate and unbiased."

During the course of her first project—spectral classification of some 1,100 bright stars from the southern sky—Cannon developed a new system. Characteristic of her cautious approach, the method followed Pickering's alphabetical scheme much more closely than had Maury's more elaborate one. However, Cannon made several important modifications. Working with Pick-

Director Edward Pickering *(right)* led the Harvard College Observatory to a turn-of-the-century golden age, thanks in part to the efforts of a staff of women, who received from twenty-five to thirty-five cents an hour for their labors. Shown below in a 1925 photograph taken well after Pickering's death, the women included Antonia Maury *(third from left)*, developer of a scheme later used to identify supergiants and fast-spinning stars by their spectra, and Annie Jump Cannon *(fifth from left)*, who classified more than a quarter of a million stars according to a spectroscopic system she refined.

ering, she rearranged the formerly alphabetical sequence to make a smoother progression in the lines marking several elements other than hydrogen. This arrangement also had the unintended result of organizing stars sequentially by color, from blue-white through yellow to red. The main spectral classes now ran from O to B, A, F, G, K, and M. Some anonymous genius later converted that sequence into the mnemonic entreaty that became familiar to generations of astronomy students: Oh, Be A Fine Girl, Kiss Me. She also combined or omitted some groups and added decimal subdivisions for several of the letters, giving the scheme a new precision. The Sun, for example, now fell in class G2, meaning it was two-tenths of the way between types G and K.

Although Cannon's sequence, like its predecessors, was based on the empirical study of spectral patterns, it also turned out to constitute a scale of stellar temperatures. By the early 1900s, laboratory studies of spectra were leading astronomers to suspect that the color of hot objects was directly linked to their surface temperature, and that those colors followed the order of the spectrum. The hottest stars were blue-white, emitting the most energy in short-wavelength blue light; the coolest radiated most of their energy in long-wavelength red light. From later research, astronomers learned to use this information to determine the temperature of a star by comparing the intensity of light at two carefully chosen wavelengths in the star's spectrum, measured with photoelectric detectors. Scientists found a wide range of temperatures in normal stars, now known to run from a high of 50,000 degrees Kelvin in the hot O stars to less than 3,000 degrees in the cool M variety.

Cannon's comparatively simple yet versatile scheme, known thereafter as the Henry Draper classification system, became the basis for the most colossal study of its kind ever undertaken. Pickering had long wanted to launch a comprehensive survey of the sky, classifying the spectra of all stars brighter than the ninth or tenth magnitude, some fifty times fainter than the eye can see. (Magnitude, as described on pages 22-23, is a measure of the brightness of a star, with the smallest numbers representing the brightest.) The survey took in nearly a quarter of a million stars, and the new system was the clear-cut choice for classifying them.

Cannon began the survey on October 11, 1911, quickly working out an efficient procedure for handling each photographic plate. Because the spectra were often no more than faint blurs, she held the transparent plate over a mirror that reflected skylight and examined the spectrum through a magnifying glass. Taking in each pattern of lines virtually at a glance, she instantly recognized the star's classification group and called out its combination of letters and numbers. An assistant would record it in a notebook while Cannon took up the next plate. In this manner, she stepped up the efficiency of the classification process until she was sometimes handling plates at the phenomenal rate of more than three per minute. On September 30, 1915, not four years after starting her monumental task, she completed it.

The resulting *Henry Draper Catalogue*, published in nine volumes beginning in 1918, contained the spectra of 225,300 stars. Together with later

supplements describing more than 130,000 additional stars examined by Cannon, the catalog became the standard in the field, as did her classification system. Cannon's achievement soon earned her awards and honorary degrees from other institutions, but it was not until 1938, when she was nearing the age of seventy-five, that Harvard granted her academic status by naming her professor of astronomy. Formal recognition came even later for Antonia Maury. In 1943 three American astronomers—William Morgan, Philip Keenan, and Edith Kellman—incorporated her idea of classification subdivisions into a revised version of the Draper scheme known as the MKK system. The same year, the American Astronomical Society awarded Maury its Annie J. Cannon prize for developing the system that Edward Pickering had discarded nearly a half-century earlier as too complicated.

The prodigious undertakings of Fleming, Maury, and Cannon in describing and organizing hundreds of thousands of stars formed the backbone of modern astronomy. Their efforts made it possible to see connections between the observed characteristics of stars—connections that would be revealed shortly after the turn of the century in an ingenious chart developed by two astronomers on opposite sides of the Atlantic.

PUTTING THE PIECES TOGETHER

The chart owed its inspiration in large part to the underappreciated contribution of Antonia Maury. Intrigued by the relative width and darkness of certain lines in the spectra she studied, Maury had found that these characteristics varied even among stars of the same color and spectral class, hence the subdivisions a, b, and c. These subdivisions—especially c, which denoted a few rare stars with very dark, narrow lines—eventually came under the scrutiny of Danish astronomer Ejnar Hertzsprung.

Hertzsprung was not an academically trained astronomer. His father, a government bureaucrat who had himself studied astronomy but did not believe it would provide a decent living, advised his son to pursue chemical engineering in college. But an interest in the chemistry of photography led Hertzsprung to stellar photography and thence to astronomy in 1902, when he was already nearly thirty. Teaching himself the trade by cadging time on telescopes at observatories in Copenhagen, the late bloomer began publishing the results of his research as early as 1905, albeit in a journal of scientific photography that was rarely read by astronomers.

The Dane's studies led him to suspect that the intrinsic brightness of stars might parallel their spectra and temperatures, with the cool red stars also being the dimmest. When he examined stars in loose stellar clusters such as the Hyades—whose absolute magnitudes relative to one another can be estimated because the stars are roughly equidistant from Earth—he found that this general correlation seemed to hold. But the stars in Maury's c subdivision were a glaring exception. Unlike one kind of relatively common and nearby red star,

Categorized according to Annie Jump Cannon's OBAFGKM system, strips of film containing stellar spectra *(right)* illustrate the patterns typical of each class of star from B *(top)* to M. Perhaps because of their rarity, no O stars are included in the display, which was published by the Harvard College Observatory in about 1900. Spectral class, later found to correspond closely to a star's surface temperature, forms the third column of the pages pictured below, taken from one of Cannon's 125 catalog notebooks. The notes also record a brightness-related value called intensity (abbreviated "Int"), location, and identifying numbers for each star.

which was dim as well as cool, Maury's red c stars, much more distant and rare, were intrinsically very bright. Hertzsprung, realizing that he had found two distinct types of stars, called the groups dwarfs and giants.

In 1908 he wrote to Harvard informing Pickering of the significance of Maury's subdivisions and urging him to incorporate them into the Henry Draper classification system. Pickering ignored the advice, believing that subtle variations in the lines might result from inadequacies in the equipment. Hertzsprung wrote again, more emphatically. Calling Maury's work "the most important advancement in stellar classification" since the pioneering work of Father Secchi, he declared: "To neglect the c-properties in classifying stellar spectra, I think, is nearly the same as if the zoologist, who has detected the deciding differences between a whale and a fish, would continue to classify them together." Pickering did not change his mind.

When Hertzsprung plotted his findings on a diagram, in which one axis represented color and the other absolute magnitude, the results showed bright bluish stars at the left and dim reddish stars at right. Most stars fell into a narrow band that Hertzsprung called the main sequence. The exceptions, red giants, formed a separate group.

Hertzsprung's breakthrough work became widely known only because at about that time an established American astronomer was independently laboring on similar problems. The American, Henry Norris Russell, had all the professional credentials Hertzsprung lacked. Russell was the son of a prominent Long Island Presbyterian minister and had graduated from Princeton in 1897 with the highest marks in that institution's history. He then studied physics and mathematics at Cambridge University before returning to Princeton to teach. Tall and lean, he was always the center of attention, whether at social gatherings, where he loved acting out charades or reciting verse from memory, or in the classroom, where his forceful lectures held generations of students spellbound. Russell would contribute so much to knowledge of the universe that one day he would be dubbed "dean of American astronomers."

Like the Dane whose work he did not yet know, the American was interested in finding the relationship, if any, between the temperature and brightness of stars. Russell knew that stars of the same temperature (as determined by their spectra) give off equal amounts of radiation per square mile of surface; therefore, if a dim star and a bright star are at the same distance from Earth and have the same temperature, the dim star must simply be smaller. Using methods similar to those of Hertzsprung, Russell found the distance to a number of stars, whose absolute magnitudes could then be determined. He then compared their intrinsic brightness to their spectra—and found the same two classes of stars that Hertzsprung had discovered. Like his counterpart across the sea, Russell also determined that the two types of red stars were giants and dwarfs in size as well as brightness.

In 1910 Russell learned of Hertzsprung's discoveries from Karl

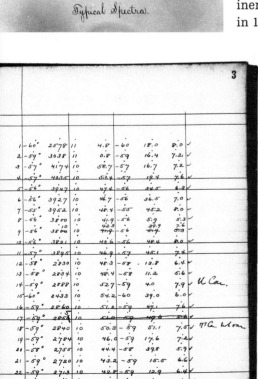

Typical Spectra.

SORTING THE STARS

In the early 1900s, astronomers Ejnar Hertzsprung and Henry Norris Russell independently found that when they plotted stellar luminosity against spectral class, a categorization that reflects a star's color and temperature, the stars fell into distinct groups. At the time, both astronomers believed that all stars followed a similar evolutionary cycle. Russell in particular thought the Hertzsprung-Russell, or H-R, diagram, as the chart came to be known, showed stars caught at different stages in their lives as they moved leftward from upper right onto a diagonal band Hertzsprung had named the main sequence and then down the sequence to lower right. Within decades, the new discipline of nuclear physics proved these speculations faulty, but the observations recorded on the H-R diagram continue to drive much of the research in stellar evolution.

As it turns out, Russell was partly right in thinking that the diagram was a kind of snapshot of stars at different points in their development. However, modern theorists hold that individual stars rarely travel either up or down the main sequence. Rather, a given star enters the sequence at a fixed position dictated by its mass, which also determines the timing of its eventual movement off the sequence.

Because of the physics of stellar fusion *(pages 51-53)*, the most massive stars are both very hot and very bright; their bluish color and the patterns of their spectral lines mark them as class O and B stars. In the course of their rapid evolution, the O and B stars exhaust their store of fuel and move off the diagonal band, swelling into giants or supergiants, as represented in the upper right corner of the diagram. Intermediate-mass yellow stars like the Sun occupy the center of the main sequence. Cooler and fainter than O and B stars, they consume their fuel at a slower pace and remain on the main sequence for billions of years before growing into giants and then dwindling to white dwarfs. The stars that remain almost indefinitely on the main sequence are cold, dim M stars, whose low mass makes them extremely conservative consumers of stellar fuel.

A Hertzsprung-Russell diagram is a kind of stellar census that sorts all of the known stars according to their luminosity, expressed in absolute magnitude *(pages 22-23)*, and spectral class, encompassing color and surface temperature. On the vertical axis, brighter stars appear near the top, fainter ones near the bottom; stellar temperature decreases from left to right, as indicated by background colors and the letters representing spectral class. More than 90 percent of the stars fall along the broad diagonal band called the main sequence, which associates lower surface temperature with lower luminosity. The remaining stars scatter into other quadrants: brilliant yet cool giants and supergiants at upper right, hot but dim dwarfs at lower left.

This variation on the H-R diagram reveals that for stars on the main sequence, size is associated with both temperature and luminosity. The hot, bright blue stars at upper left are also the sequence's largest; moderate yellow stars like the Sun are of correspondingly moderate size; and cold, dim red stars are the smallest bodies on the sequence. But for a minority of stars off the main sequence, luminosity is influenced more by size than by temperature. Planet-size white dwarfs, though generally hotter than the Sun, have such small surfaces that they are much fainter. Conversely, red giants and most supergiants are often no warmer than the Sun, yet their vast size—up to a thousand times the Sun's radius—renders them a million times more luminous.

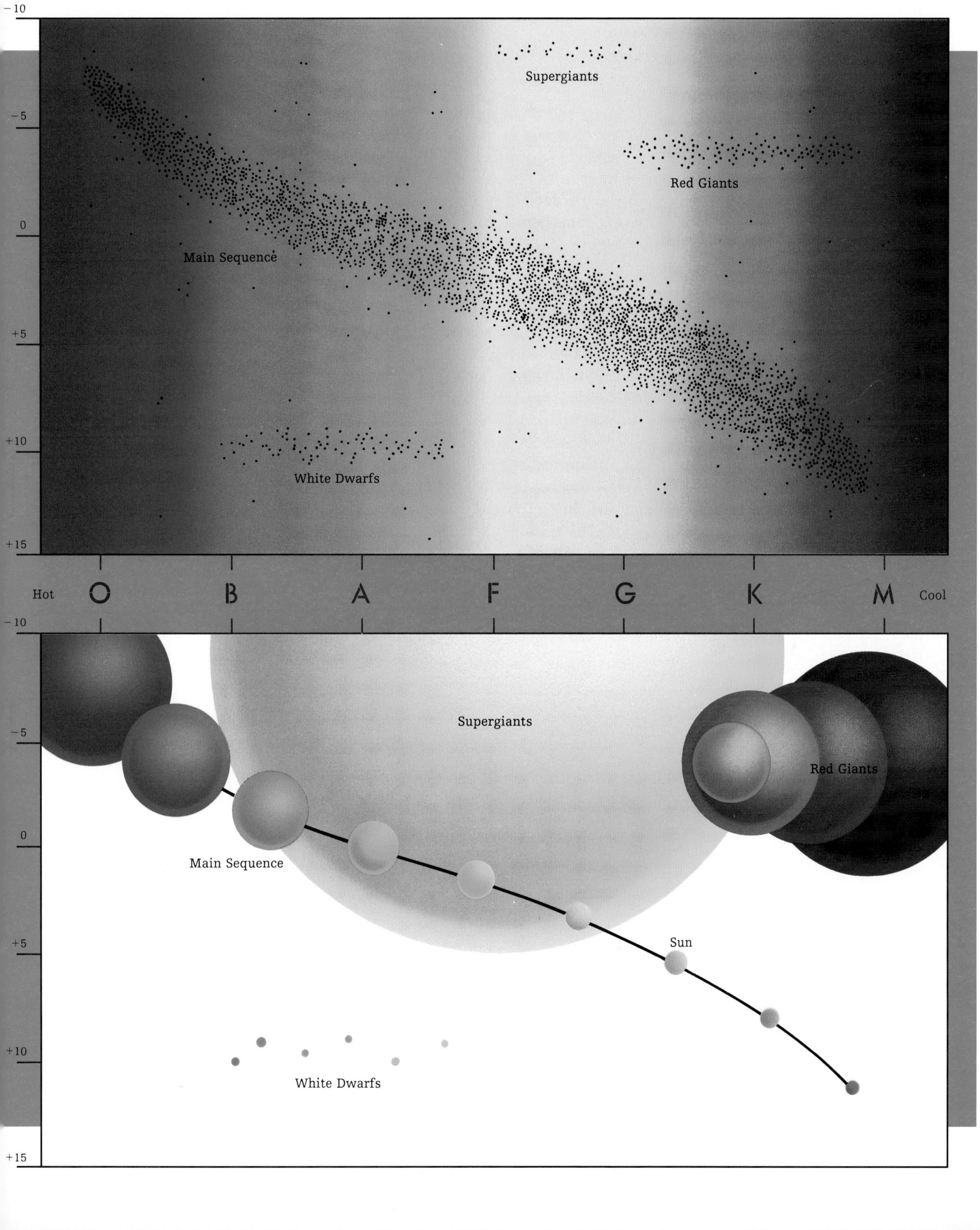

Supergiants

Red Giants

Main Sequence

White Dwarfs

−10
−5
0
+5
+10
+15

Hot O B A F G K M Cool

−10
−5
0
+5
+10
+15

Supergiants

Red Giants

Main Sequence

Sun

White Dwarfs

Schwarzschild, a noted German physicist and Hertzsprung's mentor. Hertzsprung and Russell began a cordial correspondence that seems not to have suffered when Russell attained the limelight by presenting the results of his own research at meetings of astronomers in 1913 and then publishing them the following year. The Dane, unusually modest and single-minded in his pursuit of science, was unconcerned about matters of professional reputation and priority in publishing. As noted by one of his colleagues, "His eyes were always on the stars."

To illustrate his findings, Russell published the diagram that portrayed his research in the form that would become famous. For data he used the stars whose distances he had determined; like Hertzsprung, he established absolute magnitude as one axis of his graph. However, he used the Harvard classification system, the Oh, Be A Fine Girl sequence of spectral types indicating both color and temperature, as his other axis. The results paralleled those of Hertzsprung, but the graph was rotated ninety degrees clockwise, with brightness increasing toward the top and color/temperature decreasing toward the right. As on the Hertzsprung diagram, about 90 percent of the stars fell onto a main sequence. Red giants clustered together in the upper right-hand corner. In describing his diagram, Russell acknowledged that Hertzsprung's work had preceded his own. The simple, useful graph eventually came to be known as the Hertzsprung-Russell, or H-R, diagram *(pages 30-31)*.

For Henry Norris Russell, the H-R diagram was especially important because it appeared to illustrate and confirm his view of how stars evolve. Until then, most astronomers had suspected that stellar evolution followed the spectral sequence; a star began life hot and blue-white and ended it cool and red. Such a view was implicit in the spectral classification systems constructed by Maury, Cannon, and the other Harvard researchers.

Russell, by contrast, proposed that the two types of red stars he and Hertzsprung had identified—the giants and dwarfs—represented the first and last stages of the life cycle. Every star, he theorized, began as a red giant in spectral class M. Then it shrank and heated up, before cooling and continuing to contract—in effect descending the diagonal band on the H-R diagram until the star wound up as a cool red dwarf back in class M. Like most of his colleagues, he believed that every star in the sky had the same life story, and the chart simply showed stars at different ages.

Although his diagrammatic portrait of the sky remains a classic, Russell's conclusions about stellar evolution were premature. Unfortunately, the astronomer lacked crucial information about the processes that fuel a star's development. He and his predecessors had compiled an impressive description of the exteriors of thousands of stars—their temperatures, sizes, and colors—but the scientists knew next to nothing about how a star works. Such knowledge awaited groundbreaking efforts in the fields of relativity and quantum physics that were to transform twentieth-century astronomy. In scarcely a decade, other investigators would build on the solid foundation of the classifiers to reveal the unimaginable inner world of stars.

SPECTRAL SECRETS

Light began to reveal its secrets to science in 1666, when Isaac Newton passed a beam of sunlight through a prism and saw a rainbow—the spectrum—appear on his wall. But there was more to such a fan of colors than met the eye. Nineteenth-century experimenters found dark lines in the Sun's spectrum and the spectra of stars, and bright lines in the spectra of light emitted by different gases heated to incandescence in the laboratory. Some of the various patterns of dark and bright lines matched—a discovery that unlocked the code of cosmic chemistry. As the detective work continued, astronomers found that the tenuous radiation they were studying extended beyond the familiar array of visible colors into electromagnetic regions the human eye cannot perceive. And, in this century, they learned how the stellar spectra are formed. The key lies in the dual—and not wholly definable—nature of light and all other forms of electromagnetic radiation. In some respects, the radiation behaves like a wave: It moves outward like ripples on a pond, traveling at a constant speed in all directions from its source. The distance between wave crests can range from under a billionth of a centimeter (gamma rays) to more than a thousand kilometers (very long wavelength radio waves).

In other respects, however, electromagnetic radiation behaves like a particle—a very tiny bullet, say—capable of knocking electrons out of atoms in a metal surface. The energy that performs such feats comes in individual units called photons. Each photon has a specific energy, associated with a specific wavelength. As explained on the following pages, a star's spectral imprint is produced by the interaction between photons generated by the star and the atoms in the gas composing the star's surface. From these runic patterns, students of the science of spectroscopy discern not only a star's composition but many other distinguishing traits, including its temperature and speed of rotation and the strength of its magnetic field.

Gamma Rays — X-Rays — Ultraviolet — Infrared

The electromagnetic spectrum embraces all radiation, from short-wavelength, high-energy gamma rays *(left)* through visible light to long-

THE GENESIS OF SPECTRA

A spectrum is, in effect, an illustrated record of a series of discrete energy transactions: the amount of energy gained or lost when an electron is disturbed. Electrons orbit the nucleus of an atom much as moons do a planet, but unlike moons, they can occupy only certain orbits, called energy levels. Unless it is jostled by a passing photon or neighboring atom, an electron will remain in its lowest possible level, or ground state. To jump to a higher level—an excited state—an electron must absorb exactly the amount of energy required to raise it to that level. Radiation passing through a cloud of atoms whose electrons are mostly in the ground state can thus lose many photons of a given energy, or wavelength. The result is a gap—or absorption line—at that wavelength in the spectrum.

Excited states are usually of brief duration, and electrons tend to drop back down, seeking the ground state. When an electron does this, it emits energy in the form of a photon whose wavelength is equal to the energy difference between the starting and ending levels. Because the intervals between energy levels in atoms of a given element are the same whether the electrons are emitting or absorbing photons, the emission and absorption line patterns are identical.

Very crowded, energetic circumstances produce a third type of spectrum—one of uninterrupted color. Such spectra commonly result when energy levels are greatly disturbed by extremely vigorous atomic collisions, changing the energy needed to shift electrons from one level to another. Because this smearing happens in a multitude of atoms, a broad, continuous band of energy emissions is produced. If an electron is detached from an atom altogether, it may contribute to such a continuous spectrum; no longer bound to discrete energy levels, it can absorb and reradiate photons of any wavelength. When many such electrons are bumped by passing photons, atoms, or other particles, they emit photons of many wavelengths.

As illustrated at right with helium atoms, which have two electrons apiece, the hotter or denser the gas, the more excited the electrons and the more the gas itself will glow, resulting in a spectrum showing either emission lines or—under very dense conditions—a

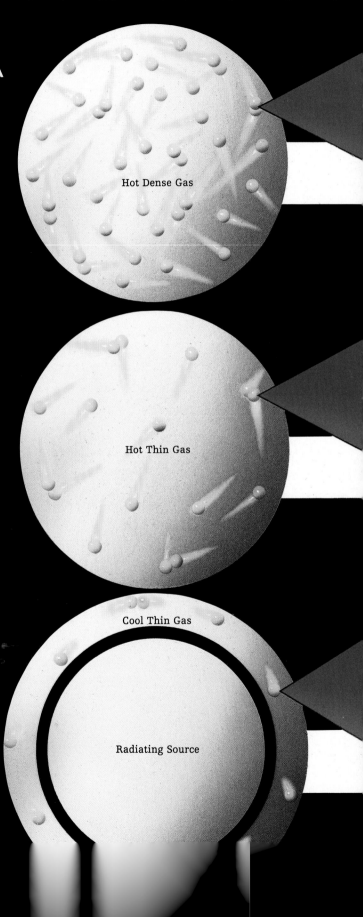

Hot Dense Gas

Hot Thin Gas

Cool Thin Gas

Radiating Source

wavelength, low-energy radio waves *(right)*. The photons making up this radiation all travel at 186,000 miles per second in a va

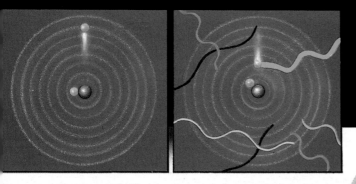

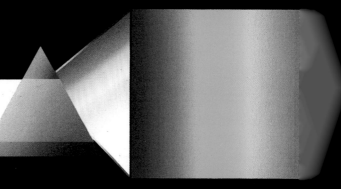

Continuous spectrum. In the hot, dense gas that occurs deep inside a star, vigorous collisions among atoms smear the distinctions between energy levels *(left)*. Atoms collide, raising electrons to higher levels *(above, left)*; in dropping back, each electron emits at a wavelength determined by the energy level distortions in its particular atom *(above, right)*. The product of countless atoms is light at all wavelengths *(right)*.

Light of all wavelengths has all

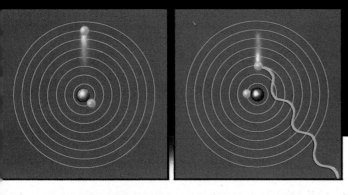

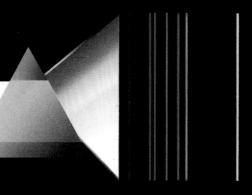

Emission spectrum. In hot gas at lower densities, such as the gas of a star's outer atmosphere *(left)*, collisions between atoms can knock an electron to a well-defined higher level *(above, left)*, but within milliseconds it drops back *(above, right)*, giving up a photon whose wavelength equals the energy difference between the two levels. The total effect of all atoms of a given chemical element is a distinctive pattern *(right)* of emitted wavelengths.

Helium emits specific wavelengt

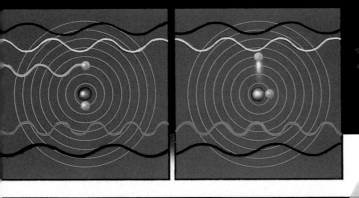

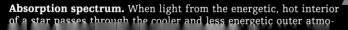

Absorption spectrum. When light from the energetic, hot interior of a star passes through the cooler and less energetic outer atmo-

PROFILE OF A STAR

Given the typical temperatures and densities of the gases in stars, stellar spectra are generally characterized by absorption rather than emission lines. Far from being identical, however, absorption spectra are as individual and revealing as portraits. Depending on the condition of the gases making up the star, the pattern of lines may be simple or complex, and the lines themselves can range in strength from pale to black.

Whether, and to what degree, an element makes its presence known in an absorption spectrum is largely a function of the star's surface temperature. At temperatures below 3,000 degrees Kelvin, the electrons in

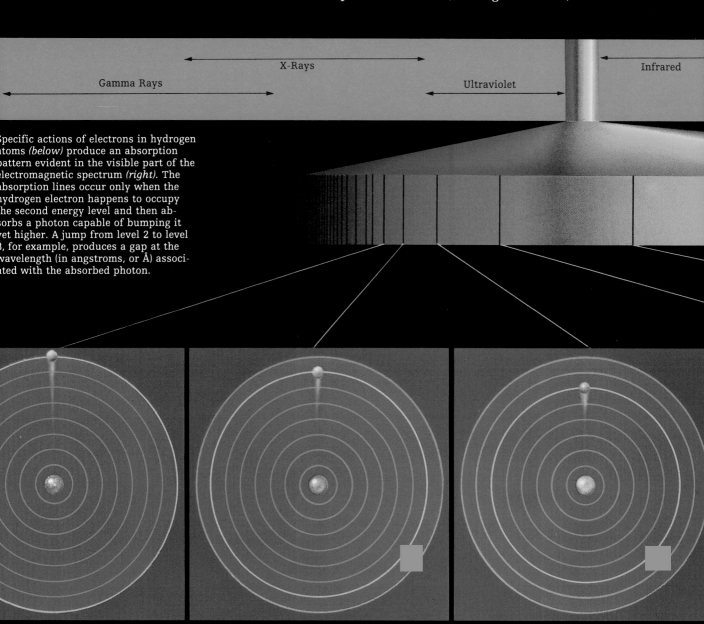

Specific actions of electrons in hydrogen atoms *(below)* produce an absorption pattern evident in the visible part of the electromagnetic spectrum *(right)*. The absorption lines occur only when the hydrogen electron happens to occupy the second energy level and then absorbs a photon capable of bumping it yet higher. A jump from level 2 to level 3, for example, produces a gap at the wavelength (in angstroms, or Å) associated with the absorbed photon.

Gamma Rays

X-Rays

Ultraviolet

Infrared

atoms of hydrogen, for example, are almost never bumped out of the ground state. Photons that might be absorbed by such electrons are associated with wavelengths beyond the violet end of the visible spectrum, and no absorption lines appear. As shown below, hydrogen lines become visible when they are produced by electrons that have been bumped by energetic collisions to the second energy level. Then, further absorptions remove photons of less energetic wavelengths.

With rising temperatures, collisions become more frequent, and the number of atoms with electrons occupying the second orbit increases. Not only do the lines become visible, they also become darker—signifying that more and more photons of associated wavelengths are being absorbed. Above 10,000 degrees Kelvin, however, the collisions become so violent that hydrogen electrons are sometimes ripped free of their atoms. With fewer intact atoms, the strength of the absorption lines decreases.

The absorption line pattern of each element follows a similar temperature-related rise and fall, with an optimal temperature at which the lines are strongest. Astronomers can thus deduce a star's temperature from the relative strengths of its absorption lines.

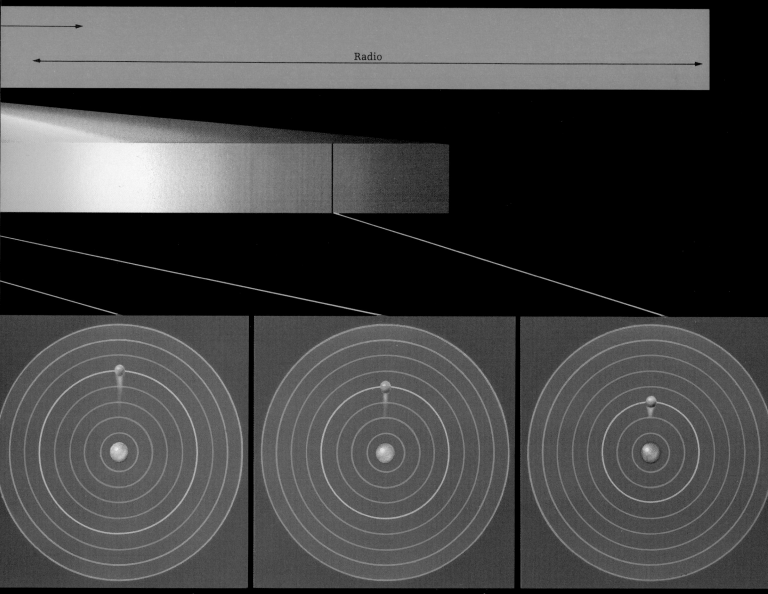

Level 2 to 5: violet line at 4,340 Å   Level 2 to 4: blue-green line at 4,861 Å   Level 2 to 3: red line at 6,563 Å

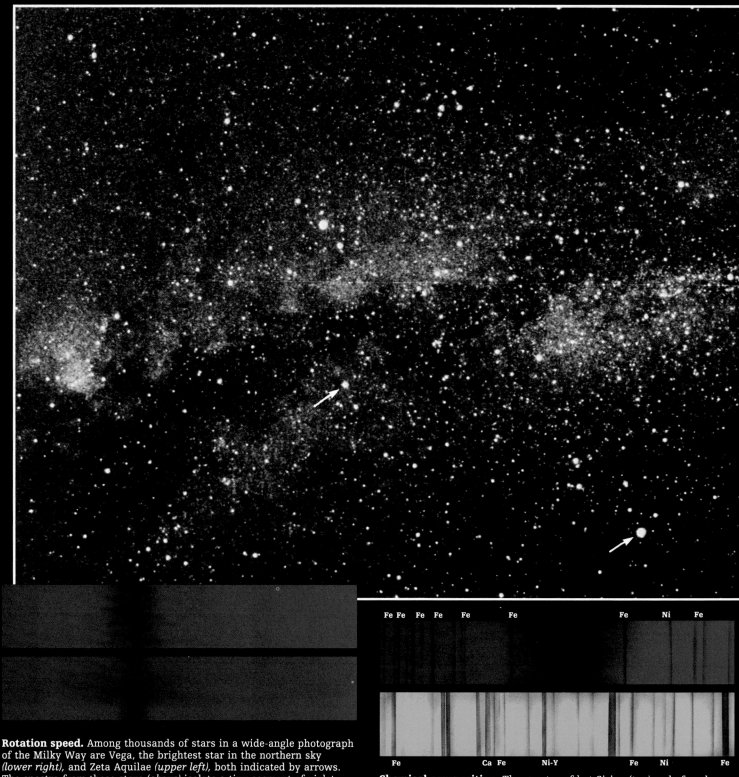

Fe Fe Fe Fe   Fe   Fe   Fe   Ni   Fe

Fe   Ca Fe   Ni-Y   Fe   Ni   Fe

Rotation speed. Among thousands of stars in a wide-angle photograph of the Milky Way are Vega, the brightest star in the northern sky *(lower right),* and Zeta Aquilae *(upper left),* both indicated by arrows. The spectra from these stars *(above)* isolate a tiny segment of violet light around a strong hydrogen absorption line to show the effect of stellar rotation. Vega *(top),* spinning at a stately fifteen kilometers per second, shows a line that is sharp and clear. But Zeta Aquilae rotates twenty-three times faster, and the same hydrogen line is broadened by the Doppler effect.

Chemical composition. The spectra of hot Sirius *(top)* and of cooler Arcturus display myriad absorption lines, indicating that the stars are composed of a variety of elements, including iron (Fe), calcium (Ca), and a mix of nickel (Ni) and the rare element yttrium (Y). Other lines in Sirius's spectrum are mostly from titanium, and the broad feature is the signature of hydrogen.

A Stellar Lineup

Temperature is only one of many things astronomers can read in the array of lines in a star's visible spectrum. The strengths and widths of the dark gaps are products of not only the chemical composition of the star's atmosphere but also such factors as pressure, rotation, and magnetism.

Simply identifying the chemical origins of the lines in a given spectrum is a painstaking process. Iron alone can produce hundreds of lines, and a mixture of elements will result in overlapping patterns. Nevertheless, by carefully comparing stellar spectra with those produced in the laboratory, astronomers have gained a certain facility at pattern recognition.

One of the more useful tools in analyzing stellar spectra is the phenomenon known as the Doppler effect. If the source of light is moving toward or away from the viewer, the wavelengths will compress or stretch, thus displacing absorption lines toward the violet or the red end of the spectrum, respectively.

The eight spectra that are shown on these two pages reveal some of the ways that scientists use the ephemeral gleam of starlight to reveal the nature of the galaxy's radiant inhabitants.

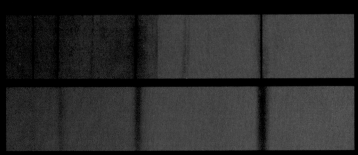

Density. Spectral lines may also be widened by pressure as a result of more frequent atomic collisions in a dense gas, which smear the energy levels. The strong lines of hydrogen are narrow in the spectrum of the huge, tenuous supergiant HR 1040 *(top).* But in the smaller, dense dwarf star Theta Virginis, they are dramatically broadened. Careful measurement distinguishes this effect from rotation, since rotation affects all of a star's lines equally, whereas broadening caused by pressure varies from line to line.

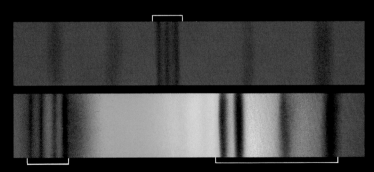

Magnetic field strength. A telltale sign of stellar magnetic fields is a spectrum that includes split absorption lines. In the spectrum of HD 215441 *(top),* a star with a magnetic field 60,000 times greater than Earth's, a tiny segment of a spectrum shows a slightly divided chromium line *(bracket).* In the spectrum of PG 1533-057, a white dwarf with a magnetic field a thousand times stronger still, both the red and blue-green hydrogen lines are greatly distorted.

2/CELESTIAL FURNACES

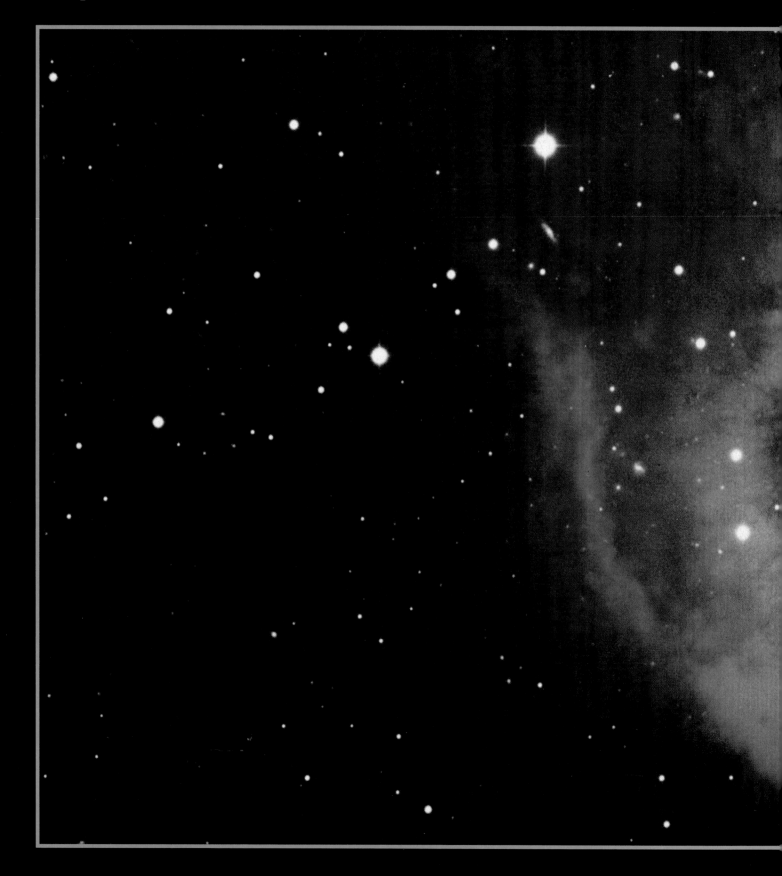

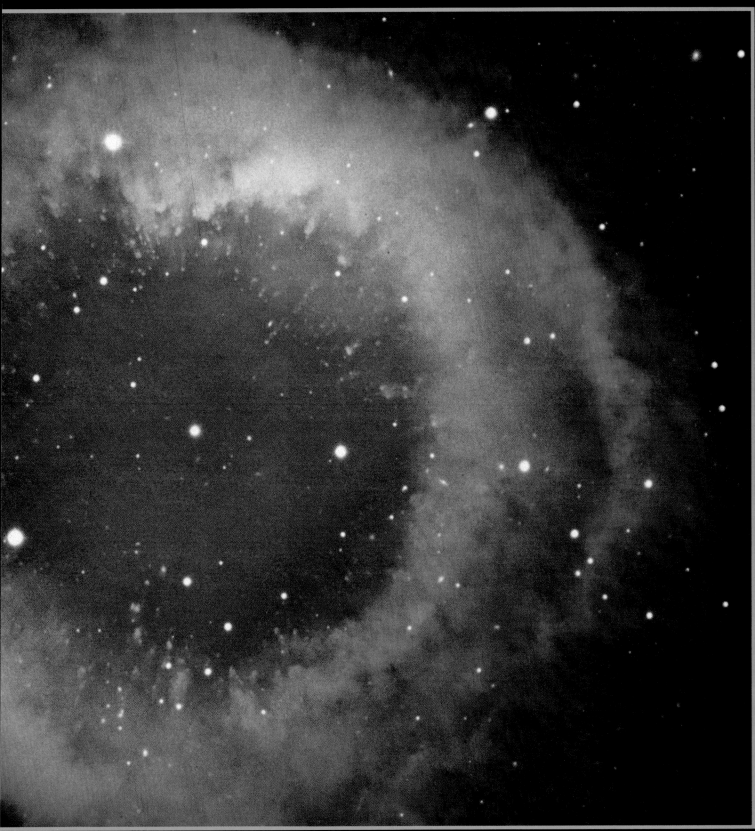

An aging star casts off its gaseous shell to form the glowing Helix nebula, called a planetary nebula because early astronomers thought such objects bore a resemblance to planetlike spheres.

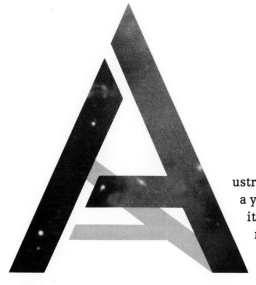ustrian physicist Fritz Houtermans was a young man intoxicated with relativity, quantum mechanics, and the mysteries of the atomic nucleus— new ideas that, in 1929, were shaking physics to its core. The old university town of Göttingen, in what is now West Germany, seethed with intellectual excitement. That March, in fact, Houtermans and a colleague had finished an important paper that applied some of those fresh ideas to astrophysics, the physics of stars.

"In the evening after we had put the finishing touches on our essay," he recalled nearly thirty years later, "I went for a walk with a pretty girl, and as it grew dark star after star lit up the sky. 'How beautifully they twinkle,' exclaimed my companion. I puffed out my chest and said, 'And since yesterday, I know why!' " Even after three decades, Houtermans was still bemused by her reaction: "She didn't seem the least impressed! Did she believe me? Apparently it was a matter of complete indifference to her!"

The young lady was probably not as impervious to astrophysics as Houtermans remembered: She later married his collaborator on that paper, British astronomer Robert Atkinson. But Houtermans's deflated feelings did not last long in any case. His paper with Atkinson, in which the two men clarified the fusion process in the center of stars, would prove to be a milestone in what was one of the most productive periods in the history of astronomy.

The years from 1910 to 1940 saw physicists embroiled in the quantum and relativity revolutions and astronomers racing to apply those new ideas in their efforts to understand the lives of stars. Their most profound discovery was that stars are brilliant not through mystery or magic but because they are nuclear furnaces, engines that liberate fierce energies by converting hydrogen into helium. The individual histories of stellar birth, life, and death became comprehensible when scientists realized that the laws of physics and chemistry governing the Earth also govern the galaxy's host of suns.

A QUESTION OF ENERGY

From the beginning of this period, astronomers were certain of one thing. They would never fathom the nature of stars until they answered three interlocking riddles: What is the internal structure of a star? How does a star change and evolve with time? And where does a star get its energy?

Of the three, the question of energy was the most baffling. It had dogged astronomers for more than fifty years, ever since they had realized it was a question at all. Until the mid-nineteenth century, most people had held to the ancient belief that the Sun and stars were made of a special material that simply had the ability to shine: It could no more stop shining than objects on Earth could stop deteriorating with time.

But that belief was dealt a deathblow in 1847, when the German physicist Hermann L. F. von Helmholtz first wrote down one of the most important principles in physics, the law of conservation of energy. He declared that energy could be neither created nor destroyed; it could only change form. Thus, the chemical energy in coal could be liberated by fire and transformed into light and heat. The heat of burning fuel could be transformed into motion and useful work by a steam engine. And the kinetic energy of a moving train could be transformed back into heat again by applying the brakes, which would become hot from friction. However, as Helmholtz himself was quick to realize, this new conservation law implied that the Sun could no longer be dismissed as luminous by nature. And yet nothing found on Earth seemed to provide a source of energy on a scale vast enough to fuel the Sun and stars. The Sun could not possibly be a burning lump of coal, for example; if it were, it would last only about 10,000 years.

In fact, the only reasonable source of energy that anyone could think of was gravity itself. Both Helmholtz and the British physicist Lord Kelvin worked out the details in the 1850s and 1860s. They suggested that the Sun and stars could make up for the energy they radiated by slowly contracting. As the cooling material shrank inward, it would be compressed and, like any gaseous material that is compressed, would grow hot and continue to radiate. Indeed, according to their calculations, the Sun would only need to shrink by some

In 1904, physicist Ernest Rutherford increased the presumed age of Earth more than tenfold. Using radioactive dating techniques, he established that a piece of uranium-filled mineral called pitchblende was 700 million years old. This discovery convinced astronomers that their estimates of the age of the Sun must be revised.

seventy feet per year to maintain its observed luminosity—a trivial amount, considering that the Sun is just under a million miles across. Furthermore, the Sun could keep on shining in this way for tens of millions of years.

Even in Kelvin and Helmholtz's day, however, astronomers were not completely comfortable with this argument. Yes, gravitational contraction might be able to power the Sun for many millions of years. But by the end of the nineteenth century, geologists and biologists were beginning to accumulate fossil evidence *(page 43)* that the Earth was perhaps as much as a billion years old. (Today the accepted age is 4.6 billion years.) So gravitational contraction was not enough. Conventional physics was not enough. Astronomers were in dire need of a new kind of energy source, one that could fuel the Sun and the stars for a very long time. And in the first decade of the twentieth century, someone appeared who would undertake to find it.

A MASTERFUL EXPLAINER

In 1913 Arthur Stanley Eddington was elected to the Plumian chair of astronomy and experimental philosophy at Cambridge University, a post from which he dominated the field of astronomy for the next thirty years. As a colleague wrote after his death in 1944, his work brought stellar astronomy to life, "infusing it with his sense of real physics and endowing it with aspects of splendid beauty. Eddington will always be our incomparable pioneer."

Privately, Eddington was shy and studious, an inveterate pipe smoker who relaxed by reading detective stories and solving crossword puzzles. Born in 1882, the son of a Quaker schoolmaster who died when Eddington was two, he had grown up with his mother and his older sister in their quiet home in the village of Weston-super-Mare, in Somerset. Eddington remained a Quaker throughout his life, and during World War I he resisted military service as a conscientious objector. He never married; in fact, when he was named director of the Cambridge Observatory in 1914, he brought his mother and sister to live with him in the director's house.

In his public life, however, Eddington displayed considerable wit and charm, and he had a disarming way with words. In a serious book on astrophysics he could write this about the inside of a star: "Try to picture the tumult! Dishevelled atoms tear along at 50 miles a second with only a few tatters left of their elaborate cloaks of electrons torn from them in the scrimmage. The lost electrons are speeding a hundred times faster to find new resting places. Look out! There is nearly a collision as an electron approaches an atomic nucleus; but putting on speed it sweeps round it in a sharp curve." To the lay public he was not just a scientist but an immensely popular explainer of science, a man whose books and lectures did more than anything else to make the new astronomy accessible to society at large.

Most important, Eddington was gifted with a vivid scientific imagination, an intuitive grasp of physical principles, and a superb mathematical ability. It was he, for example, who introduced Albert Einstein's general theory of relativity to England. Eddington's 1923 book *The Mathematical Theory of*

Astronomer and mathematician Arthur Stanley Eddington, seated here in a 1923 photograph with Albert Einstein, was one of the first scientists of his day to grasp the full implications of Einstein's theory of relativity. Calling the theory "a revolution in thought," Eddington used his understanding of it to figure out how nuclear fusion might be the source of stellar energy.

Relativity was the first comprehensive text on that subject written in English; more than thirty years later, Einstein said he still considered it the finest presentation of relativity in any language.

Eddington's efforts to understand stars began in earnest in 1916, when he turned his attention to the question of stellar structure and evolution. At that time the best available theory of stellar evolution was one proposed by Henry Norris Russell in 1913 on the basis of the diagram later named after Russell and his colleague, Ejnar Hertzsprung. While allowing for the possibility of some kind of energy generation "of a radio-active nature," Russell had assumed that the stars are basically powered by Kelvin-Helmholtz gravitational contraction. According to this scheme, when a star is a young, extended giant, its material is as tenuous and compressible as air on Earth. In technical terms, it obeys the ideal gas law: Its molecules do not interact. But when it reaches the main sequence, the star has contracted so much that its atoms actually begin to touch. At this point, the material in the star suddenly becomes stiffer—more like water than air—and cannot continue to contract as rapidly, marking the beginning of the end. Smaller, stiffer dwarf stars like the Sun continue to shrink, and gravitational contraction continues to supply energy, but not nearly as efficiently. The rest of their lives are spent slowly cooling off; ultimately they become cold cinders.

Until Eddington came along, Russell's model was widely accepted. Eddington himself believed in it, at first, because it elegantly explained the otherwise mysterious regularities of the Hertzsprung-Russell diagram in terms of physics that everyone recognized. However, Eddington realized that neither Russell nor anyone else had a very clear idea of the physical processes that were taking place inside a star. The Cambridge don set out to fill in the gaps, and eight years later, Russell's model lay in ruins. Eddington first considered the equations that governed the internal structure of a star.

Those equations simply stated what physicists and astronomers had known for many years: namely, that stars exist in a state of delicate equilibrium. Gravity pulls inward and pressure pushes outward. If the two are not in balance at every point, from the core to the surface, the star will expand or contract until it reaches a balance.

However, Eddington saw that the theorists up until that time had been neglecting one of the most important sources of pressure inside stars. Under ordinary circumstances, pressure arises from the motion of atoms in hot gas, just as it would on Earth. But if a star gets hot enough, said Eddington, radiation—primarily heat and light generated in atomic reactions—exerts a pressure. Each atom finds itself in the midst of a firestorm of energy. Eddington calculated that in the deep interior of a star, where temperatures can reach many tens of millions of degrees Kelvin, the pressure of radiation alone would amount to roughly 25 million times the pressure of the atmosphere at the Earth's surface. In fact, he concluded that if a star had more than about a hundred solar masses, radiation pressure would tear it apart.

Meanwhile, Eddington was also invoking some brand-new ideas in atomic theory to show that, contrary to the most essential feature of Russell's model, a star does not stiffen when it reaches a certain density. According to the nuclear model developed by New Zealand-born physicist Ernest Rutherford in 1911, atoms were not just solid blobs, as Russell had assumed. Rather, each atom was more like a little solar system, with a cloud of electrons whirling around a tiny, dense nucleus. A collection of such atoms in the core of a star would be ionized, or stripped of electrons, by the incredible temperature, leaving behind nuclei so far apart that even in these conditions they would not touch. The gas inside a star thus would still obey the ideal gas law—a finding that made Russell's distinction between stars on the giant branch and stars on the main sequence hopelessly wrong.

Russell's model was dead, and with it the idea that gravitational contraction was the power source of the stars. It should be said that Russell himself was both generous and graceful in accepting the demise of his theory. In a 1925 paper on stellar evolution he wrote, "Several investigators . . . have contributed to this field, but much the largest share is Eddington's."

But if gravity was not the answer, what was? Eddington was convinced that the solution must lie with Einstein's famous equation, $E = mc^2$. At a time when Einstein's theories were not so widely accepted as they are now, other astronomers were skeptical, but Eddington saw the conversion of matter into energy as the only process that could explain the prodigious outpouring of stars. For him, the only question was how the conversion took place.

Radioactivity was familiar enough in the 1920s. But none of the known decay processes, in which elements naturally break down into constituent particles, releasing radiation, were remotely energetic enough to explain the Sun. One possible solution, Eddington thought, was the conversion of hydrogen into helium in a process he called "the transmutation of elements"—known now as fusion. If one were to start with four pounds of hydrogen, he pointed out, and somehow force the protons constituting their nuclei to stick together and form helium nuclei, then the resulting mass of helium would weigh only 3.97 pounds. The difference in mass, which represents the binding energy of the constituents, would show up as heat and light.

As even Eddington recognized, transmutation seemed to have its own problems. One was fuel supply: In the early 1920s it was not yet clear that the Sun had enough hydrogen to support such a conversion. Within a few years, that objection melted away. Cecilia Payne (later Payne-Gaposchkin), a brilliant graduate student at Harvard, had used the university's huge collection of stellar spectra to survey the composition of a number of stars. To her astonishment, she found that all bright stars were essentially identical: overwhelmingly made of hydrogen, with ample amounts of helium. Together, in fact, the two elements made up 98 percent of the matter in the universe.

Less easily vanquished was the law of physics that causes positively

Pioneering English astronomer Cecilia Payne established in 1925 that hydrogen is the most abundant element in the Sun and other stars—a finding that is the foundation of current theories about the evolution of stars and galaxies.

charged protons to repel each other. According to Eddington's calculations, the temperature at the Sun's core was an almost unimaginable 40 million degrees Kelvin. (The best modern estimates are about 14 million degrees.) But according to other researchers, even that extreme was pathetically inadequate. Temperatures would have to reach tens of billions of degrees before electrostatic repulsion could be overcome. The Sun was nowhere close.

Despite this unresolved issue, Eddington made an emphatic case for transmutation in his 1926 masterpiece, *The Internal Constitution of the Stars.* Somehow, he maintained, atomic nuclei must react at temperatures far lower than most physicists believed possible. As he wrote in a popular version of his book a year later, "I am aware that many critics consider [that] the stars are not hot enough. The critics lay themselves open to an obvious retort; we tell them to go and find a hotter place."

TOWARD UNDERSTANDING FUSION

Eddington's faith was justified. Even as his book was being published, two German physicists, Werner Heisenberg and Erwin Schrödinger, were independently supplying the key to fusion with their descriptions of quantum mechanics—a new theory of elementary particles and their interactions.

It would seem in retrospect that Eddington and others had been baffled by hydrogen fusion because they had assumed that subatomic particles such as protons behave in the same way as everyday objects such as rocks or billiard balls. Given that assumption, fusion was, to put it mildly, a difficult proposition. In one useful analogy, two protons might be likened to two balls, one sitting in a deep, narrow pit and the other rolling around freely on the ground. The pit represents the powerful but extremely short-range force that binds particles together in an atomic nucleus. The two particles will fuse if the second proton can somehow roll into that pit to join the first. The problem is that the pit is surrounded by a very high, very steep hill, which corresponds to the repulsion between the two protons. Before the two can fuse, the free-rolling proton has to make its way up and over the hill. To do so, it has to be moving very fast indeed; otherwise it will only roll partway up and then back down again. Hence, physicists shared the widespread conviction that fusion could not occur except at temperatures in the tens of billions of degrees; anything less ferocious would not get the protons moving fast enough.

With the advent of quantum mechanics, however, the fusion problem was back to square one. In particular, Heisenberg and Schrödinger showed that subatomic particles are not like billiard balls. According to a tenet now known as the Heisenberg uncertainty principle, it is better to think of atoms as fuzzy clouds of unpredictability, objects that jitter and dart so insanely that no one can ever be quite sure of where they are or how fast they are moving. Furthermore, such particles do not roll around like ordinary objects; they move through space by vibrating and oscillating, as if they were waves.

This revised description of matter opened a new field of science, and in those early years no place in the world was more active in developing it than the University of Göttingen. There, the great German physicist Max Born had gathered a coterie of talented young men, including Heisenberg, the Hungarian-born Edward Teller, and a Russian émigré physicist named George Gamow. Gamow's investigations yielded the first hint as to how quantum mechanics might resolve astronomy's impasse on fusion.

Gamow started by looking at the opposite problem: escape from nuclei. Everyone knew that certain unstable atomic nuclei—radium, for example—decay by emitting so-called alpha particles (another name for helium nuclei). And yet everyone also knew that a radium nucleus sits in the same kind of walled pit that a proton does. So if the pit is surrounded by a wall, how does the alpha particle get out? Gamow's answer, published in 1928, was a new quantum phenomenon he called the tunnel effect. Imagine an alpha particle rattling around in the nuclear pit, he said. Because it behaves like a cloud of uncertainty and not like a billiard ball, it tends not to be confined in such a small space but instead to leak out. And there is a chance, said Gamow, a small but nonzero probability, that its incessant quantum jittering will cause it to punch through the surrounding wall without ever going over the top.

To most physicists this tunneling business seemed akin to magic, but there was no arguing with the fact that it worked. Gamow's calculations on radioactive alpha decay fit the data beautifully. Furthermore, the research led to another intriguing possibility, which Gamow was also the first to point out: If particles could tunnel out of a nucleus, then why could they not also tunnel in?

This was the challenge taken up by Fritz Houtermans and Robert Atkinson, two more of Göttingen's bright young men. In March 1929, with Gamow's encouragement, they set about applying his tunneling ideas to the problem of stellar energy production and quickly showed that Eddington's faith in fusion was justified. Even at temperatures as "low" as 40 million degrees Kelvin, protons could tunnel through their electrostatic mountain ranges, find each other, and react—not often, but often enough. Indeed, the two men were able to calculate the rate of fusion reactions among a whole range of nuclei. Although they could not tell precisely which reactions might be occurring in the Sun, they did verify that hydrogen was the key: Only when hydrogen was part of the reaction would enough energy be released.

And that explains Houtermans's exhilaration on the spring evening when he went walking with a female companion. Not only had he and Atkinson found the energy source of the stars, they had laid the foundations for the whole theory of thermonuclear fusion reactions. (They submitted their article to the journal *Zeitschrift für Physik* with the whimsical title "How to Cook a Helium Nucleus in the Potential Pot." The unwhimsical editor made them change it to something more prosaic.)

Radical and promising as it was, quantum tunneling was still

not the whole answer. How did the fusion process work in detail? Exactly what reactions were involved? What was the precise rate of energy production? And for that matter, how could four protons manage to find each other and stick together simultaneously to form a helium nucleus? Houtermans and Atkinson, hampered by the crudeness of their data, had not been able to say, and neither would anyone else for nearly ten years.

The ten-year research gap was only partly caused by the need for better data. In the 1930s, European physics was thrown into chaos by the rise of Nazism. As many of the best and brightest scientists fled for their lives, Germany—so recently the intellectual center for the revolutions in relativity and quantum mechanics—was left a scientific wasteland.

Among the many who found a new home in the United States were George Gamow and Edward Teller. Settling in Washington, D.C., in 1934 and 1935, respectively, they briefly made George Washington University a leading center for theoretical astrophysics before moving on to other research positions in the next two decades. By the late 1930s they were convinced that the problem of stellar energy production was ripe for a solution. In the spring of 1938, they therefore arranged a small scientific meeting in Washington to bring physicists and astronomers together on the problem.

ON THE VERGE OF A SOLUTION

The meeting was more successful than they could have hoped—not least because one of the thirty-four attendees was another young refugee from Germany: the nuclear physicist Hans Bethe, who had settled in Upstate New York at Cornell University. Bethe's interest in stellar energy production had been piqued by a student of Gamow's and Teller's, Charles L. Critchfield, who was pursuing a problem in nuclear physics. Just the year before, the Baron Carl Friedrich von Weizsäcker, a physicist who had remained in Germany, had proposed that the solution to the solar energy problem lay with what is now known as the proton-proton reaction. As the name suggests, this series of nuclear reactions within a star's core starts with two protons (each the nucleus of a hydrogen atom). What follows is a complex sequence of events whose net result is that four protons are lost, while one helium nucleus, containing two protons and two neutrons, is created—along with a lot of energy (pages 51-53).

Critchfield wanted to calculate the rate at which this reaction produced energy, so that he could see if it agreed with the measured energy output of the Sun: about 380 sextillion kilowatts. To complete the calculation, however, he had to understand the quantum-mechanical intricacies of the nuclei involved. So he turned to Bethe for help.

"By the time of the Washington conference we had our calculations done," Bethe later recalled. And by combining those calculations with some new and more accurate estimates of the temperature at the center of the Sun, he and Critchfield were able to show that von Weizsäcker's theory fit the data beautifully.

Physicist Hans Bethe (below, right), here receiving the 1967 Nobel prize for physics from King Gustav VI Adolf of Sweden, helped to unravel the mystery of stellar energy production by elucidating a fusion process known as the carbon cycle (pages 52-53).

However, there was still a problem. The proton-proton reaction explained the energy output of the Sun, but it fell far short of the energy radiated by hotter, more massive stars. "So I said to myself, 'Well, maybe there is something else for these bigger stars,' " recalled Bethe.

Bethe mulled over the problem on his way back to Cornell. Another kind of reaction chain was clearly needed. Upon his return he began a systematic search through laboratory data on nuclear reactions, starting from the lightest and simplest nuclei and working his way through the heavier species, checking every one that could possibly react with protons. "Nothing seemed to work," he recalled, "and I was almost ready to give up. But when I tried carbon, it worked. So, you see, this was a discovery by persistence, and not by brains."

Bethe realized that each carbon nucleus in the center of the Sun acts rather like a soft ball of clay: Each time a proton hits it, the proton sticks. The nucleus collects four extra protons in this way before the particles split off as a single helium nucleus. The result once again is four protons gone, one helium nucleus created, and a lot of energy generated.

According to Bethe's figures, the carbon cycle would contribute very little energy in a star the size of the Sun, which is why the proton-proton reaction explains the Sun so well. But with more and more mass, the carbon cycle would soon become a star's dominant energy source. Sirius, for example, with a mass more than twice that of the Sun, derives its power almost entirely from the carbon cycle.

Working out the details took Bethe six weeks. In the summer of 1938, he sent the finished paper to the *Physical Review*. But then a graduate student, Robert Marshak, pointed out to Bethe that the New York Academy of Sciences was offering a $500 prize for the best unpublished paper about energy production in stars. "So I asked the *Physical Review* to return the paper to me, and I promised Marshak a finder's fee of ten per cent," recalled Bethe. "I got the prize and he got the fifty dollars. I used part of the prize to help my mother emigrate. The Nazis were quite willing to let her out, but they wanted $250, in dollars, to release her furniture. Part of the prize money went to liberate my mother's furniture." It was not the last prize Bethe was to win for his summer's labor. In 1967 he was awarded the Nobel prize in physics for elucidating the energy source of the stars.

STARS IN OLD AGE

Thus, by the eve of World War II the energy source of stars on the main sequence of the H-R diagram was well understood, thanks to the work of Atkinson, Houtermans, Bethe, and their colleagues. Astronomers were even beginning to understand how stars evolve—except for one nagging question. If stars shine by burning up their nuclear fuel, then what occurs at the end of a star's life when its fuel is exhausted?

As it happens, the question had been answered in the late 1930s. Few had recognized it, because the solution was opposed and ridiculed by a figure of

Fueling the Thermonuclear Fire

In the fierce heart of a star, the ancient dream of alchemy comes true. But instead of changing ordinary metals into gold, stellar transmutations produce something of infinitely greater value—vast quantities of energy. The name of the cosmic magic trick is nuclear fusion. Unlike the interactions between atoms that take place in the chemistry of everyday fuel burning, fusion occurs on a subatomic level, between the nuclei of atoms.

An atomic nucleus may contain two types of particles *(below)*—positively charged protons and primal forms of protons called neutrons, which have no charge. (Electrons, negatively charged particles that orbit the nucleus, much as planets orbit the Sun, are minor players in the fusion process.) Elements are distinguished by the precise number of protons in their nuclei: Hydrogen, for example, has one proton, helium two, lithium three, carbon six. The number of neutrons in atoms of a given element is not fixed, and

these slight variations in neutron count give rise to different versions of each element, known as isotopes.

At the raging temperatures found in the stellar furnace, atoms are stripped of their electrons, leaving behind bare nuclei. Under nonstellar conditions, these nuclei, containing positively charged protons, would never fuse; electrical repulsion would act to keep like-charged particles apart. But the furious heat in the stellar core imparts enough energy to move some nuclei to within a tenth of a trillionth of an inch of one another—a critical distance near enough for the short-range nuclear effect known as the strong force to take over, causing fusion. A reverse process also takes place: Atoms may spontaneously change, or decay, into isotopes of the same element or even into other elements. As shown here and on the next two pages, the strong force drives the nuclei together to form new elements in an endless cycle of fusion and decay that varies with the temperature of the star.

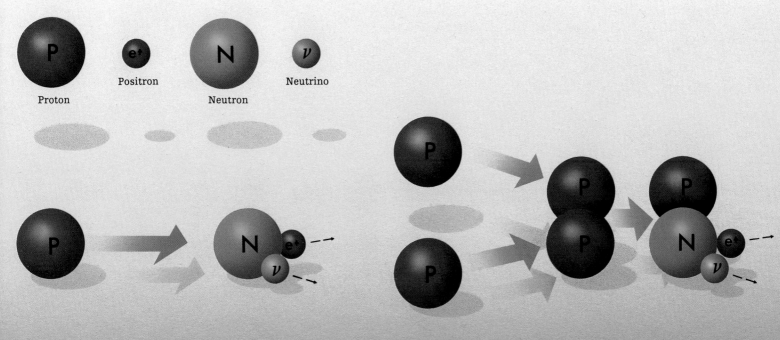

Among the particles involved in fusion *(top)* is the proton, whose positive charge is caused by a tiny particle called a positron. When a proton decays *(above)*, it yields a positron, a neutron, and a tiny, almost massless neutrino.

In the simplest fusion reaction, two protons combine. One quickly decays into a neutron, a positron, and a neutrino. The positron and neutrino carry off the energy released in the reaction, leaving behind a hydrogen isotope, known as a deuteron, made of one proton and one neutron.

FUSION VARIATIONS

Fusion reactions vary in the ingredients they start and end with, and different chains—three of which are shown here—tend to dominate at different temperatures in the star's interior. In stars like the Sun, with cores at less than 15 million degrees Kelvin, the dom-inant chain reaction is known as the proton-proton reaction, which starts with the simple two-proton fusion described on page 51. In hotter stars, with cores up to 100 million degrees, the so-called carbon cycle predominates. As stars age and their center temperatures soar above 100 million degrees, a chain known as the triple-alpha process fuses three helium nuclei, or alpha particles, to form a carbon nucleus.

THE PROTON-PROTON REACTION

1 The intense heat in a star's core drives two hydrogen nuclei, with one proton each, close enough together to overcome their mutual repulsion. The strong force takes over and the protons fuse.

2 One proton decays, releasing a positron and a neutrino and reverting to its primal form, the neutron. With the other proton, the neutron forms a deuteron, the nucleus of the hydrogen isotope deuterium. Another one-proton hydrogen nucleus begins to fuse.

3 The deuteron and the proton fuse to form an isotope of helium with two protons and one neutron. A similar helium nucleus approaches.

4 The two helium isotopes fuse to produce a helium nucleus with two protons and two neutrons. The helium nucleus and the two extra protons break apart with such violence that they carry away one-half of the total fusion energy.

THE CARBON CYCLE

1 In stars that are hotter than 15 million degrees Kelvin, the dominant fusion reaction involves trace amounts of carbon. Here, a carbon nucleus with six protons and six neutrons is joined by a single-proton hydrogen nucleus.

7 Instead of yielding a nucleus with eight protons and eight neutrons, the fusion expels a helium nucleus with such rapidity that its energy is the primary source of heat. A carbon nucleus remains to start the cycle over.

6 One proton in the oxygen nucleus decays into the trio of neutron, positron, and neutrino, leaving a nucleus that is an isotope of nitrogen. The addition of another hydrogen proton is imminent.

5 The fusion of the nitrogen and hydrogen nuclei yields an isotope of oxygen with eight protons and seven neutrons.

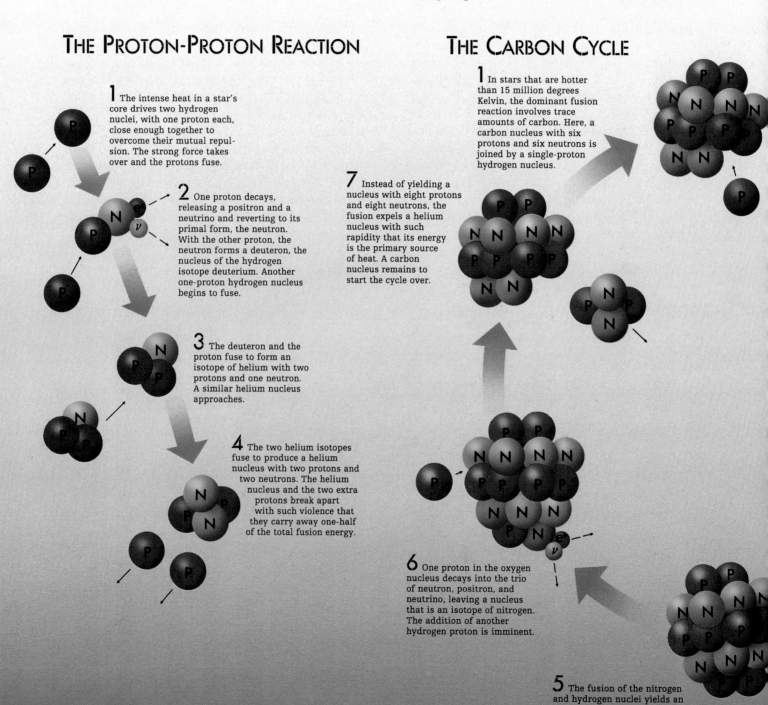

All fusion reactions have one major outcome in common: They produce vast quantities of energy. The mass of the elements produced in each reaction turns out to be a fraction less (.007 less in the proton-proton reaction, according to Einstein) than the total mass of the matter that went into the reaction. The missing mass is released as energy, carried away by photons and different types of particles, depending on the stage of the reaction. For example, in the proton-proton reaction, positrons and neutrinos do the job in one phase, protons in another; in both the carbon cycle and the triple-alpha process, energy is released throughout as short-wavelength gamma rays. The energy is absorbed by surrounding stellar matter, heating the gas and producing enough radiance to keep the Sun and stars shining for billions of years.

THE TRIPLE-ALPHA PROCESS

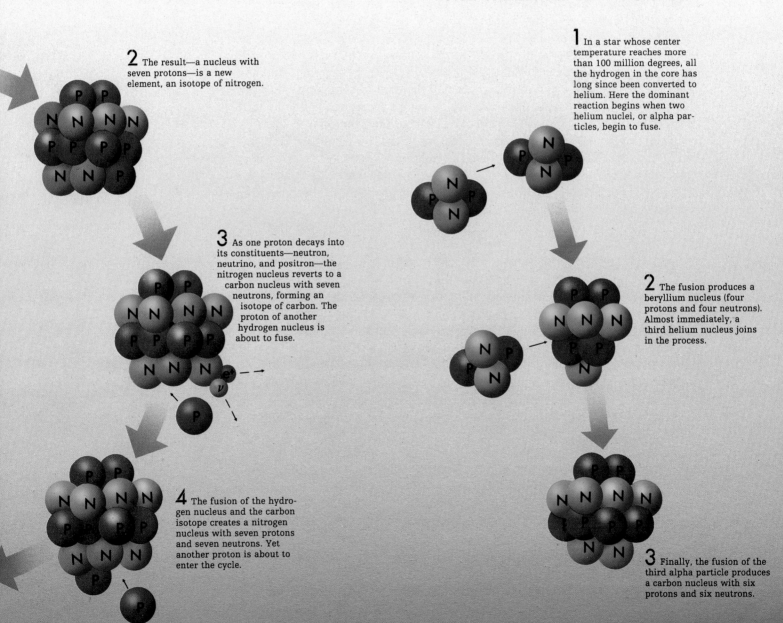

2 The result—a nucleus with seven protons—is a new element, an isotope of nitrogen.

3 As one proton decays into its constituents—neutron, neutrino, and positron—the nitrogen nucleus reverts to a carbon nucleus with seven neutrons, forming an isotope of carbon. The proton of another hydrogen nucleus is about to fuse.

4 The fusion of the hydrogen nucleus and the carbon isotope creates a nitrogen nucleus with seven protons and seven neutrons. Yet another proton is about to enter the cycle.

1 In a star whose center temperature reaches more than 100 million degrees, all the hydrogen in the core has long since been converted to helium. Here the dominant reaction begins when two helium nuclei, or alpha particles, begin to fuse.

2 The fusion produces a beryllium nucleus (four protons and four neutrons). Almost immediately, a third helium nucleus joins in the process.

3 Finally, the fusion of the third alpha particle produces a carbon nucleus with six protons and six neutrons.

enormous stature and influence, a man who had so often championed un-popular ideas himself: Arthur Stanley Eddington.

The question of stellar death turned on white dwarf stars, one of the more remarkable specimens in the astrophysical zoo. Their existence was fore-shadowed in 1844, when the German astronomer Friedrich Bessel noticed a slight irregularity in the motion of Sirius. In the next three decades, astron-omers learned the irregularity was caused by the gravitational attraction of a barely visible companion star, Sirius B, which is almost as massive as the Sun. However, Sirius B (and a handful of other such stars discovered by Eddington's day) is much smaller than the Sun. In fact, it is roughly the size of the Earth, which means that it has a fabulously high density. As Eddington wrote later, "The message of the Companion of Sirius when it was decoded ran: 'I am composed of material 3,000 times denser than anything you have come across; a ton of my material would be a little nugget that you could put in a matchbox.' What reply can one make to such a message? The reply that most of us made in 1914 was—'Shut up. Don't talk nonsense.'"

A white dwarf seemed to be an impossibility. On the one hand, said Ed-dington, a star of such incredible density would have to consist of atoms that had been completely stripped of their protective cloud of electrons and then crammed together, nucleus to nucleus. This was conceivable only if the star was extremely hot—hot enough to ionize the atoms completely. White dwarfs were in fact known to have surface temperatures of 10,000 degrees Kelvin or more. This was twice as hot as the Sun, implying interior heat on the order of 50 to 100 million degrees.

On the other hand, Eddington continued, as hot things radiate energy, they are supposed to cool off. So when a white dwarf's sea of ionized atoms cool off enough, the electrons should return to their orbits, forcing the nuclei apart as atoms fill out to normal size. If the atoms moved apart, then the white dwarf as a whole would have to expand. Paradoxically, however, to do so, it would have to press out against its own ferocious gravity, and with no more fuel to burn, it simply would not have enough energy. "I do not see how a star that had once got into this compressed condition is ever going to get out of it," wrote Eddington. "Imagine a body continually losing heat but with in-sufficient energy to grow cold."

The solution to the paradox came from quantum mechanics, which was first applied to the white dwarf problem in 1926 by Eddington's Cam-bridge colleague Ralph Howard Fowler. According to Fowler, atoms in a white dwarf are ionized, all right; the mistake lay in assuming that elec-trons would recombine with nuclei as the star cooled. The very fact that the material was so tightly packed meant that the electrons were sharply constrained in position, unable to travel very far. But according to the Heisenberg uncertainty principle, this would just force the electrons to be highly unconstrained in momentum—so unconstrained, in fact, that they would usually be moving much too fast to reunite with their nuclei. So the star would never swell up after all.

Further support for white dwarfs, said Fowler, came from the exclusion principle, a statement formulated several years earlier by the German physicist Wolfgang Pauli. The principle says that two identical particles cannot occupy the same quantum state in an atom; in a sense, they cannot be in the same place at the same time. In the context of white dwarfs, Pauli's principle meant that electrons would strongly resist being jammed on top of one another. The result would be a new kind of pressure, known as degeneracy pressure. This finally resolved Eddington's paradox. Not only would degeneracy pressure be strong enough to withstand the crushing gravity of a white dwarf, but it would be independent of temperature. It would persist even when the star cooled to absolute zero.

Eddington found Fowler's solution profoundly satisfying. His colleague had solved the mystery of white dwarfs while explaining what happens to ordinary stars when they exhaust their hydrogen fuel. They simply shrink into white dwarfdom, slowly radiating away their remaining heat until they become dark cinders held up by degeneracy pressure. But Fowler's theory had neglected something that Eddington of all people should have recognized—relativity. And in 1930, the year Eddington was knighted by the King, that oversight was brought forcibly to his attention with the arrival in Cambridge of a shy, intense, and dogged twenty-year-old from India.

A SELF-TAUGHT QUANTUM PHYSICIST

Subrahmanyan Chandrasekhar was born in 1910 in Lahore, in what is now Pakistan, and grew up in the southern Indian city of Madras. His uncle was the great Indian physicist Chandrasekhara Venkata Raman, who was to win the Nobel prize for his work in atomic physics in 1930. "The atmosphere of science was always at home," Chandrasekhar recalled many years later.

As a student at the Presidency College in Madras, the teenage Chandrasekhar excelled in mathematics and physics. From thousands of miles away, he caught a few glimpses of the new world of quantum mechanics. Chandrasekhar taught himself what he could, working from texts and papers that were often out of date. Soon he was publishing original research papers of his own. He also won a college contest for the best essay on the quantum theory. Since the prize was a book, he asked for one that he had seen at the college library, Eddington's *The Internal Constitution of the Stars*.

Clearly, this was no ordinary student. When he graduated in 1930, the young man won a Government of India scholarship to do graduate work at Cambridge University. In those days, a trip from India to England involved weeks of living on a boat. Chandrasekhar, wanting to pass the time as productively as possible, set himself a little problem: Use the equations from Eddington's book to find the internal structure of white dwarf stars. This he did, incorporating Fowler's insights on degeneracy pressure. However, his results led him to a new problem. At the densities that would prevail at the center of such a star, electrons would be moving so fast that their velocities would approach the speed of light. And that meant that the laws of relativity would apply. Chandrasekhar knew that Fowler had not taken relativity into account in his equations. Would anything change if the laws were applied?

By the time the ship reached England, Chandrasekhar had drafted two papers. One contained an answer so simple and yet so astonishing he was not sure he could believe it. Yes, relativity changed things. Above a certain mass, now known to be 1.44 times the mass of the Sun, the revised equations simply had no solution. Such massive stars would not become white dwarfs. Instead, the delicate balance between gravity and degeneracy pressure would tilt, and the star would continue to collapse inward. What then? The young scientist did not know.

More than half a century later scientists know the answer: The collapsing white dwarf would form either a neutron star or a black hole. But for Chandrasekhar in the early 1930s, the death of a massive star was an utter mystery. As he wrote in one paper, published in 1932, "Great progress in the analysis of stellar structure is not possible before we can answer the following fundamental question: . . . what happens if we go on compressing the material indefinitely?"

When he arrived in Cambridge, Chandrasekhar showed his manuscripts to Fowler, who seemed skeptical. Painfully shy and lonely, feeling very subdued by the company of intellectual giants, the student retreated to his room, emerging only for lectures and meals. But he continued to work on his theory of white dwarfs. In 1933 Chandrasekhar received his Ph.D. and was elected a fellow of Trinity College, the center of the University's scientific activity, where Eddington, Fowler, and many other luminaries were faculty members. Here, he began to feel at last that he was part of the Cambridge community. Eddington himself frequently came to visit him in his rooms, following Chandrasekhar's work almost on a daily basis; the two often dined together at the Trinity dining hall's high table, which is reserved for faculty and distinguished visitors. As Chandrasekhar's ideas about white dwarfs began to gain exposure and more people seemed to be paying attention, he felt encouraged. Thinking to answer the skeptics once and for all, he decided to work out the mathematics of white dwarfs exactly, making no assumptions or approximations.

Chandrasekhar finished that task in the autumn of 1934 and presented his results in January 1935 in a talk before the monthly meeting of the Royal Astronomical Society. His mathematical work had confirmed his earlier conclusions: If a star's mass is greater than a certain limit, it cannot end as a white dwarf but must collapse.

Speaking after Chandrasekhar, however, was Eddington, who had been listening with interest. "I do not know if I shall escape from this meeting alive," he said, "but the point of my paper is that there is no such thing as relativistic degeneracy! . . . I think there should be a law of nature to prevent a star from behaving in such an absurd way!" Eddington went on to question the way in which relativity had been combined with quantum theory—"I do not regard the offspring of such a union as born in lawful wedlock"—and to suggest that the correct formulation would eliminate the problem. As always, his talk was interrupted by appreciative laughter from the audience.

Chandrasekhar was devastated. In effect, and with no prior warning, the most prestigious and influential astronomer in the world had accused him of a fundamental conceptual error. Several people came up to him after the meeting, saying "too bad, too bad." Could Eddington be right?

In a panic, Chandrasekhar wrote to his longtime friend Leon Rosenfeld, a close colleague of Niels Bohr's in Copenhagen, asking for an authoritative statement on the issue from the physicists. Rosenfeld quickly replied that neither he nor Bohr could find any basis whatsoever for Eddington's objections. Rosenfeld forwarded Chandrasekhar's arguments to Wolfgang Pauli, and obtained a similar result: Eddington's arguments, concluded Rosenfeld, were "the wildest nonsense."

None of it disturbed Eddington. He continued to hold forth in scientific meetings, insisting that relativistic degeneracy was wrong and that there was no limit to the mass of a white dwarf. He disagreed with Chandrasekhar in public and in long private discussions. The argument ran unabated for four years, until Chandrasekhar, in effect, gave up.

"I had to make a decision," he recalled. "Am I going to continue for the rest of my life fighting . . . or change to another area of interest? I said, well, I will write a book and then change my interest. So I did." Published in the year 1939, the book was titled *An Introduction to the Study of Stellar Structure.* Nearly half a century later it is widely regarded as a classic, a work that defined the subject for years thereafter.

There is little doubt now that Chandrasekhar was right. Just as the young man had feared, Eddington's opposition in the 1930s left the majority of astronomers confused and doubtful about the subject. Almost two decades passed before Chandrasekhar's conclusions were generally accepted. By now astronomers have determined the properties of hundreds of white dwarfs, all of which fall under the Chandrasekhar limit. And the theory that an insecure twenty-year-old student began to formulate on that long-ago boat journey is now a corner-

Astrophysicist Subrahmanyan Chandrasekhar was a twenty-two-year-old graduate student when he used relativity theory to discover the maximum possible mass for a white dwarf, the last stage of a typical star's evolution. Ridiculed at first for this insight, Chandrasekhar was later vindicated as astronomers came to recognize that all known white dwarfs fall within what is now called the Chandrasekhar limit.

stone of astrophysics. For his groundbreaking work in those years, Chandrasekhar won the 1983 Nobel prize in physics.

To this day, no one really knows why Eddington was so adamantly opposed to Chandrasekhar's insight. Surprising as it may seem, the disagreement did nothing to alter the two men's friendship. They continued a warm correspondence, even after Chandrasekhar moved to the University of Chicago in 1938. In 1944, after Eddington's death, Chandrasekhar said in his eulogy, "I believe that anyone who has known Eddington will agree that he was a man of the highest integrity and character. I do not believe, for example, that he ever thought harshly of anyone. That was why it was always so easy to disagree with him on scientific matters. You could always be certain that he would never misjudge you or think ill of you on that account."

MASS IS DESTINY

With recognition of the Chandrasekhar limit, the theoretical foundation for understanding the lives of stars was complete. Moreover, astronomers came to realize that everything that distinguished one star from another—temperature, luminosity, size, life span—was determined almost entirely by one key factor: a star's mass. Variations in mass explained the observations that so misled Russell and his contemporaries. The main sequence is not a singular evolutionary pathway, as they thought; it is simply a portrait of the sky at one moment, depicting different stars of varying masses.

Astronomers with modern instruments and with modern computers now build on this foundation to give us a remarkably detailed chronicle of how stars are born, how they live, and how they die. In effect, they have replaced Russell's elegant 1913 theory of stellar evolution with a new version—more complex, perhaps, but much more realistic.

The modern theory describes how stars are born in vast clouds of interstellar hydrogen and helium, where the densest clumps of gas slowly contract under the force of their own gravity. Once a newborn star achieves stability, it takes its place on the main sequence according to its mass and stays there, barely changing until old age. Only when a star begins to exhaust its hydrogen fuel does it finally leave the main sequence, moving quickly up to the giant branch of the Hertzsprung-Russell diagram and briefly shining as a red giant, a bloated, cool, but very luminous object roughly the size of Mars's orbit.

The destiny of dying stars is guided by Chandrasekhar's 1.44 solar mass limit. After the red giant phase, most stars will have shrunk below that limit, becoming white dwarfs. Stars that start life with more than about eight times the mass of the Sun will retain enough matter in old age to stay above the dividing line. Although their fate is still debated, astronomers do know that at least some of them, too massive to settle quietly into senescence, die quickly and violently in spectacular explosions, known as supernovae. In the mid-twentieth century, these cosmic blasts became the subject of intense scrutiny as physicists sought to understand not only stellar dynamics but the origin of most of the elements in the universe.

THE LIVES OF STARS

ecades of observation combined with the revelations of nuclear theory have allowed modern astronomers to make out the simple scheme underlying the stellar universe. Each kind of star—and there are hundreds of types—represents a temporary phase in a standard life cycle. With a few adjustments, this cycle applies to every star known. All stars, for example, begin as protostars, concentrations of luminous gas found within far larger and more diffuse clouds of dust and gas. Collapsing inward under its own gravity, a protostar heats and compresses its core until hydrogen-fusion reactions ignite. At this point, the star is considered to be on the main sequence, a reference to the observed concentration of most stars on a diagonal track, or sequence, within the Hertzsprung-Russell diagram of stellar properties *(pages 30-31)*. Many stars remain on the main sequence for billions of years. But for each, there comes a time when its hydrogen supply runs out, causing the star to undergo further evolution.

A star's mass controls the onset of this crisis. Low-mass stars, for example, have correspondingly low gravity, which allows them to fuse hydrogen very slowly and stay on the main sequence almost indefinitely; high-mass stars have such high gravity, and thus such rapid reactions, that they consume their own much greater hydrogen stocks within a few tens of millions of years. After the hydrogen is gone, mass dictates how each star changes. The smallest simply consume the dregs of their fuel and wink out. Mid-size stars like the Sun go through a bewildering variety of changes, including a high-energy helium flash *(pages 64-65)*, before turning to white dwarfs. The most massive stars rush through an intricate series of fusion reactions before suffering a final spectacular collapse. The twists and turns of stellar development are chronicled on the pages that follow.

FROM OUT OF DARKNESS

Coalescing within the gas and dust of interstellar clouds, nascent stars are huge, continuously shrinking balls of hydrogen, helium, and other elements *(right)*. These gaseous spheres are much larger than mature stars; glowing with the energy of their gravitational contraction, most of them easily surpass the Sun in luminosity. Yet the dust of their surroundings renders them invisible to optical telescopes; only radio and infrared instruments can penetrate the veil.

The gas and dust clumps that give rise to protostars are themselves fragments of huge nebulous bodies called giant molecular clouds. Dozens of light-years across and composed primarily of molecules of hydrogen and other, more complex compounds, the clouds may yield thousands of protostars apiece over several million years. The component gaseous clumps in each cloud are within thirty degrees Kelvin of absolute zero—a theoretical limit at which molecules and atoms are motionless. (Commonly used by astronomers, the Kelvin scale employs degrees equivalent to Celsius degrees, but sets its zero-degree mark at absolute zero.) In a kind of cosmic irony, this deathly chill actually contributes to the molecular clouds' fecundity: Dust and gas can accumulate, since they are not being dispersed by heat, allowing the gravity of dense regions to attract material from the surrounding medium until they create protostars.

Once formed, a protostar contracts steadily for millennia before reaching the high density and million-degree temperatures needed to start nuclear fusion at its core. Except in very low mass protostars, which may never attain these conditions, fusion reactions act to halt further contraction, balancing inward gravitational pull with heat-driven expansion. When this equilibrium is reached, a star is born.

Three sizes of protostars *(red)* emerge within the cold, lumpy fragments *(gray)* of a single giant molecular cloud. Low-mass protostars arise mainly in the coolest regions *(upper left),* with temperatures around ten degrees Kelvin. High-mass varieties have sufficient gravity to form where gaseous clumps may warm to twenty degrees *(center)* and be more turbulent as a result. Mid-size protostars may appear at both temperatures.

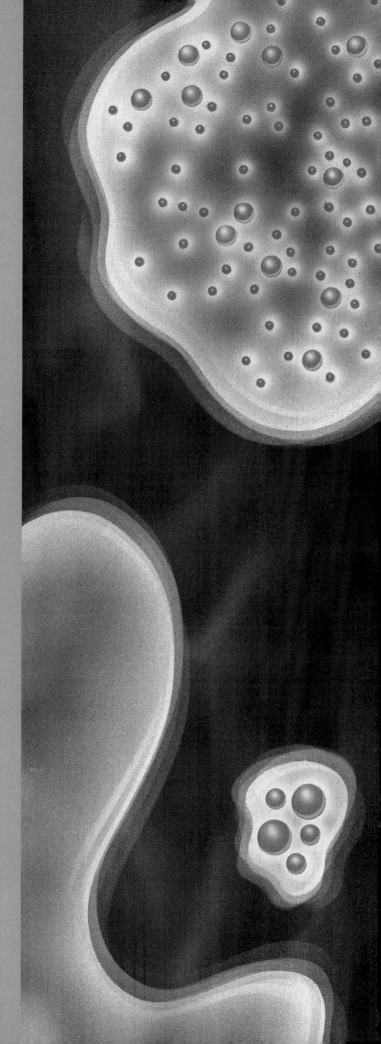

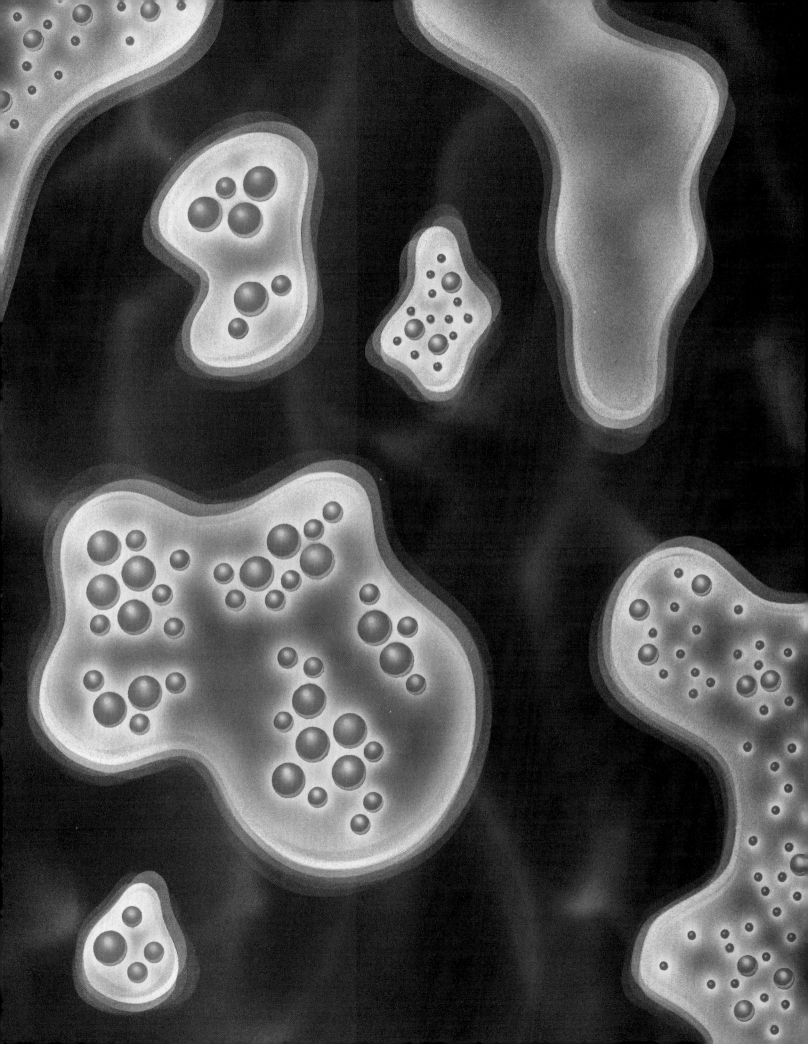

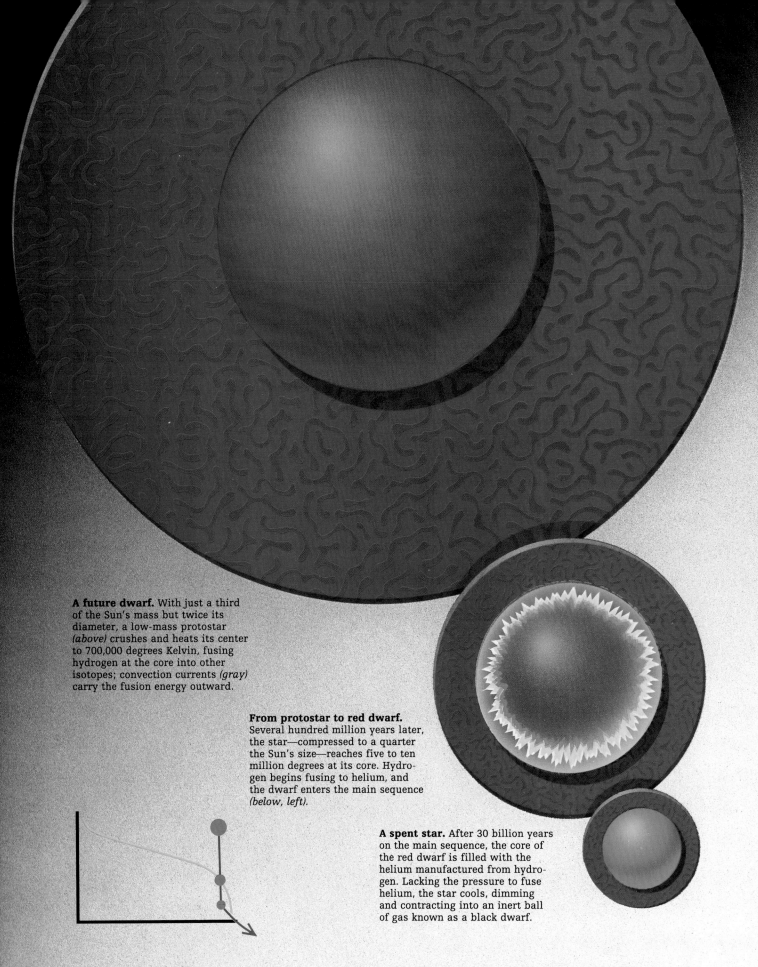

A future dwarf. With just a third
of the Sun's mass but twice its
diameter, a low-mass protostar
(above) crushes and heats its center
to 700,000 degrees Kelvin, fusing
hydrogen at the core into other
isotopes; convection currents *(gray)*
carry the fusion energy outward.

From protostar to red dwarf.
Several hundred million years later,
the star—compressed to a quarter
the Sun's size—reaches five to ten
million degrees at its core. Hydrogen begins fusing to helium, and
the dwarf enters the main sequence
(below, left).

A spent star. After 30 billion years
on the main sequence, the core of
the red dwarf is filled with the
helium manufactured from hydrogen. Lacking the pressure to fuse
helium, the star cools, dimming
and contracting into an inert ball
of gas known as a black dwarf.

62

A Long-Lived Race of Dwarfs

Most protostars are small, low-mass bodies that evolve into even smaller stars, known as dwarfs. With less than a third the mass of the Sun, they proceed slowly through very simple life cycles, shown here in separate steps keyed to the Hertzsprung-Russell diagrams at lower left of both pages. (Such diagrams chart stars' surface temperature horizontally and luminosity vertically; most stars lie on the diagonal slope of the so-called main sequence.) Two kinds of low-mass stars are known, both named for the hues characteristic of their low temperatures.

The larger variety, with at least a tenth of the Sun's mass, are red dwarfs *(opposite page),* cool stars thousands of times fainter than the Sun. Red dwarfs convert hydrogen to helium much as the Sun does, melding hydrogen nuclei at their cores into a series of variations, or isotopes, and then fusing the isotopes into helium. A red dwarf's low gravity drives the fusion process very slowly, leaving it on the main sequence far longer than any other star. A star with half of the Sun's mass, for example, could survive 200 billion years, about fifteen times the present age of the universe, and astronomers calculate that even-smaller red dwarfs will number their main sequence years in the trillions. Once the dwarf does use up its central hydrogen reservoir, scientists theorize, fusion will end. The entire star will contract further, squeezing its helium core until only the mutual repulsion of the core's electrons prevents total gravitational collapse—a standoff that physicists refer to as degeneracy. The star's heat will then dissipate, leaving behind a dark, cold sphere of compressed gas—a yet-to-be-observed object known as a black dwarf.

Less massive even than red dwarfs are brown dwarfs *(near left)*—faint bodies with masses only a few times that of the planet Jupiter. (Astronomers have yet to confirm observations of these objects.) Brown dwarfs are heated solely by their contraction and never reach the conditions required to produce helium. Instead, these failed stars fade from view within a few hundred million years, probably becoming black dwarfs just as red dwarfs do.

A stillborn star. A protostar the size of the Sun but less than a tenth its mass ripples with convection currents from its 700,000-degree core. The core's hydrogen fuses into isotopes, but its temperature and pressure are too low to fuse the isotopes to helium.

A brown dwarf. Without the steady heat supplied by hydrogen-helium fusion, a brown dwarf misses the main sequence *(below),* cooling and shrinking to the size of the Earth with its nuclear fuel still untapped.

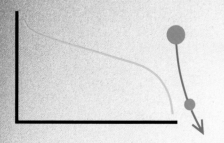

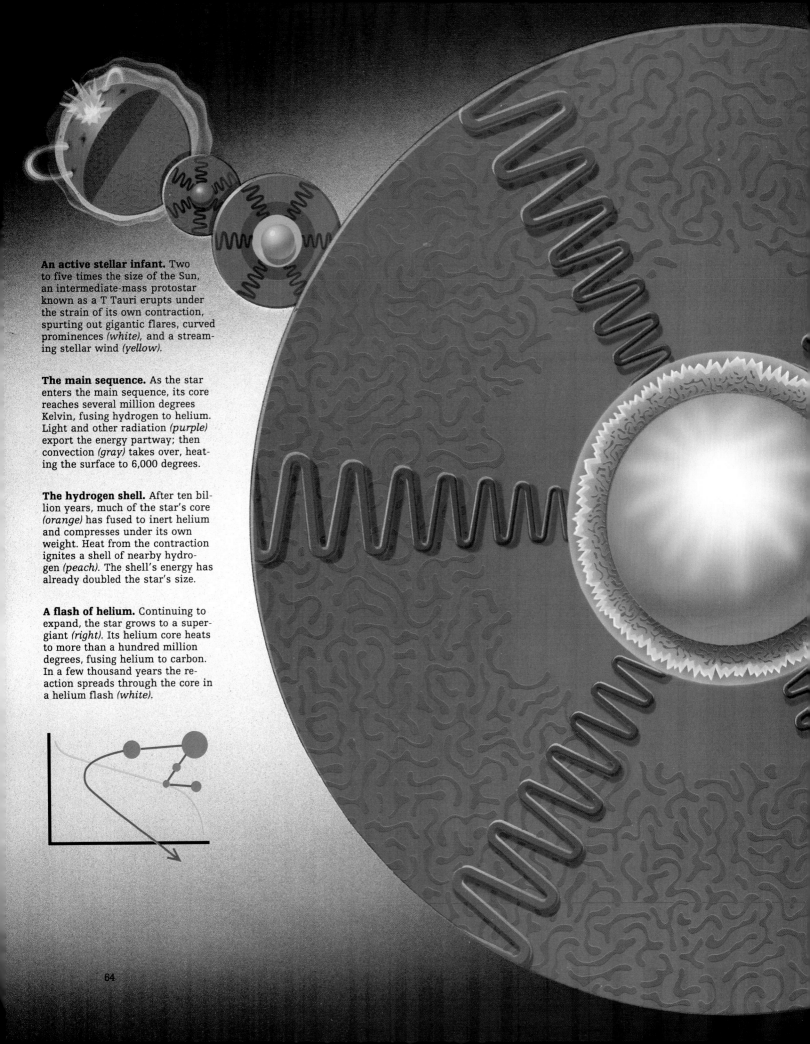

An active stellar infant. Two to five times the size of the Sun, an intermediate-mass protostar known as a T Tauri erupts under the strain of its own contraction, spurting out gigantic flares, curved prominences *(white)*, and a streaming stellar wind *(yellow)*.

The main sequence. As the star enters the main sequence, its core reaches several million degrees Kelvin, fusing hydrogen to helium. Light and other radiation *(purple)* export the energy partway; then convection *(gray)* takes over, heating the surface to 6,000 degrees.

The hydrogen shell. After ten billion years, much of the star's core *(orange)* has fused to inert helium and compresses under its own weight. Heat from the contraction ignites a shell of nearby hydrogen *(peach)*. The shell's energy has already doubled the star's size.

A flash of helium. Continuing to expand, the star grows to a supergiant *(right)*. Its helium core heats to more than a hundred million degrees, fusing helium to carbon. In a few thousand years the reaction spreads through the core in a helium flash *(white)*.

SUN-CLASS STARS

Mid-size stars like the Sun—though puny compared to true stellar giants *(pages 66-69)*—are still massive enough to escape the dead-end fate of dwarf stars, which fade out once their cores fill with helium. Sun-class stars edge past that crisis and move on to a series of further transformations. Such stars leave the main sequence in less than 50 billion years, even before exhausting their core hydrogen. A star exits once its high gravity ignites a hydrogen shell outside the still-fusing core, yielding stellar properties that no longer fit the sequence. When core fusion does end, the shell's heat not only prevents stellar collapse but puffs out the star to a hundred times its former size.

For several hundred million years, only the shell keeps the star alight. Then the helium core heats enough to fuse to carbon. But since the helium is degenerate—so compressed that its electrons' thermal motion is drastically curtailed—it does not at first expand, a delay that causes a so-called helium flash to spread through the core. The flash initiates steady helium-carbon fusion *(below)*, but in less than a hundred million years that process ends too, leaving a carbon core. With insufficient gravity to fuse carbon, the star dies; its envelope drifts off, and its core becomes a hot, highly compressed white dwarf star.

An aging giant. An old red giant fuses helium to carbon at its core while fusing hydrogen to helium in an outer ring *(red)*. Between the core and outer shell lie an inner zone rich in helium *(orange)* and a middle shell low in hydrogen *(peach)*.

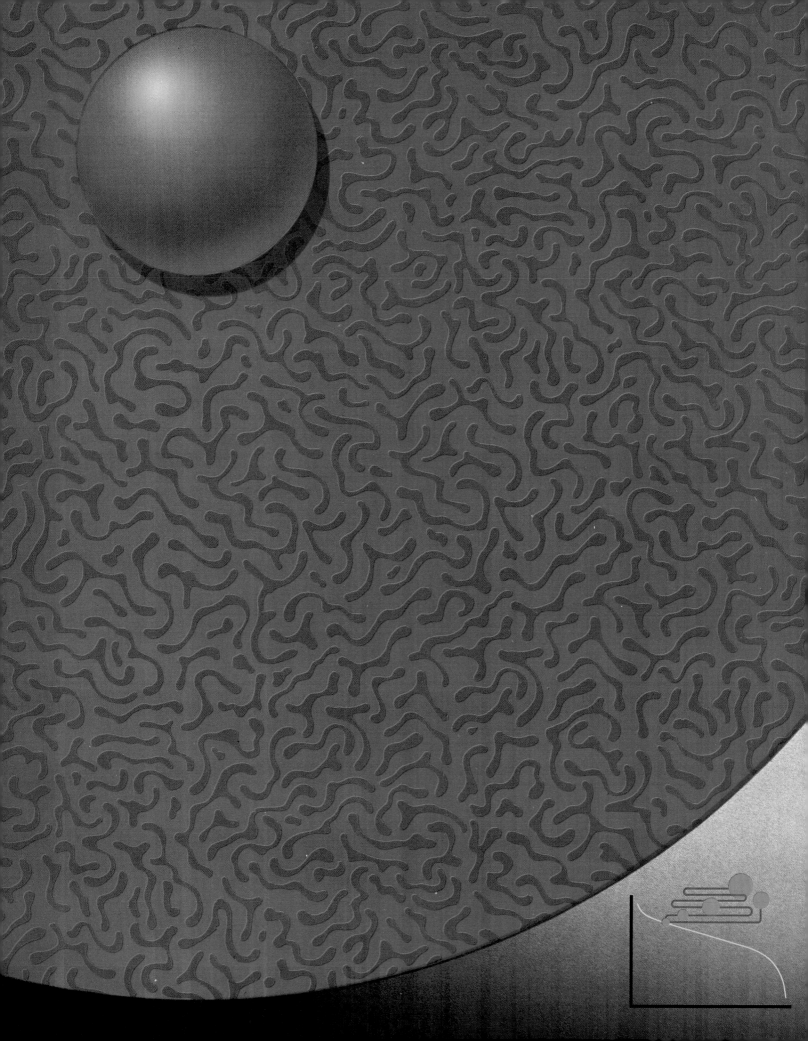

Fast-Burning Giants

Hot, bright, and blue, stars of at least three solar masses trace a quick, flamboyant course through time. Their short life spans make the large stars rare, since only those formed in the last 30 million years—and not all of those—still exist. Their extreme youth also means massive stars are still to be found near the stars they formed with. Low-mass stars have time to disperse from their original cohort, but very massive stars do not last long enough to do the same, remaining in so-called associations that are suffused with leftover wisps of gas and dust.

At first they pass quickly through much the same phases as an intermediate star *(pages 64-65)*, but massive stars have such hot cores that they change hydrogen to helium in a different way, using traces of carbon, nitrogen, and oxygen. After their cores turn to helium, the stars' enormous gravity allows fusion to continue, converting helium to carbon, carbon to neon, neon to oxygen, oxygen to silicon, and finally, silicon to iron. At this point, because iron will not fuse, a massive star's core collapses rapidly, either shrinking into a black hole *(pages 123-135)* or recoiling into a supernova explosion *(pages 89-101)*.

A huge protostar. In the course of a few thousand years, a protostar hundreds of times the size of the Sun contracts enough to approach stellar status. Convection transports heat from the core outward through a huge hydrogen envelope to the surface, itself a furnace at 3,000 degrees Kelvin.

A brief maturity. As the star fuses its core hydrogen to helium, it joins the main sequence *(below, left)*. But it remains there less than 100 million years: At six times the Sun's radius, and with four times its surface temperature, a star this size burns brightly and quickly.

A red giant. As the core changes to helium, it shrinks. Around it, a zone of depleted hydrogen *(peach)* forms, surrounded in turn by a hydrogen-rich shell *(red)*. Heated by the core, the star doubles in size—en route to supergiant status *(overleaf)*.

67

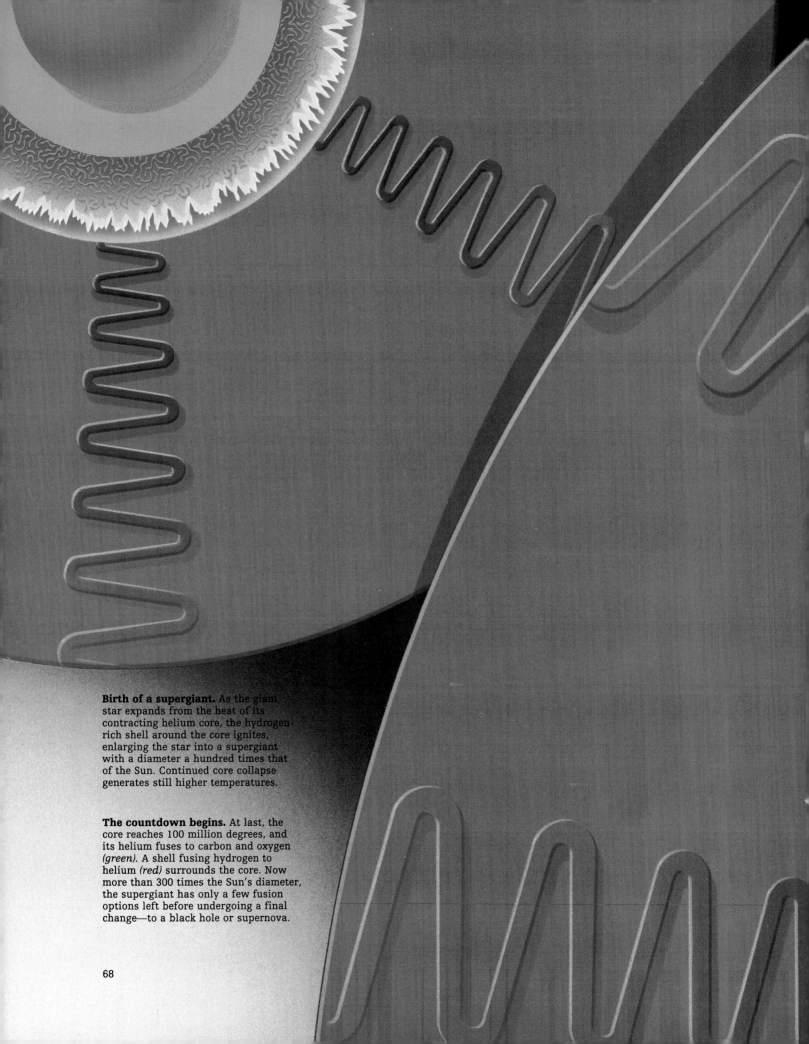

Birth of a supergiant. As the giant star expands from the heat of its contracting helium core, the hydrogen-rich shell around the core ignites, enlarging the star into a supergiant with a diameter a hundred times that of the Sun. Continued core collapse generates still higher temperatures.

The countdown begins. At last, the core reaches 100 million degrees, and its helium fuses to carbon and oxygen *(green)*. A shell fusing hydrogen to helium *(red)* surrounds the core. Now more than 300 times the Sun's diameter, the supergiant has only a few fusion options left before undergoing a final change—to a black hole or supernova.

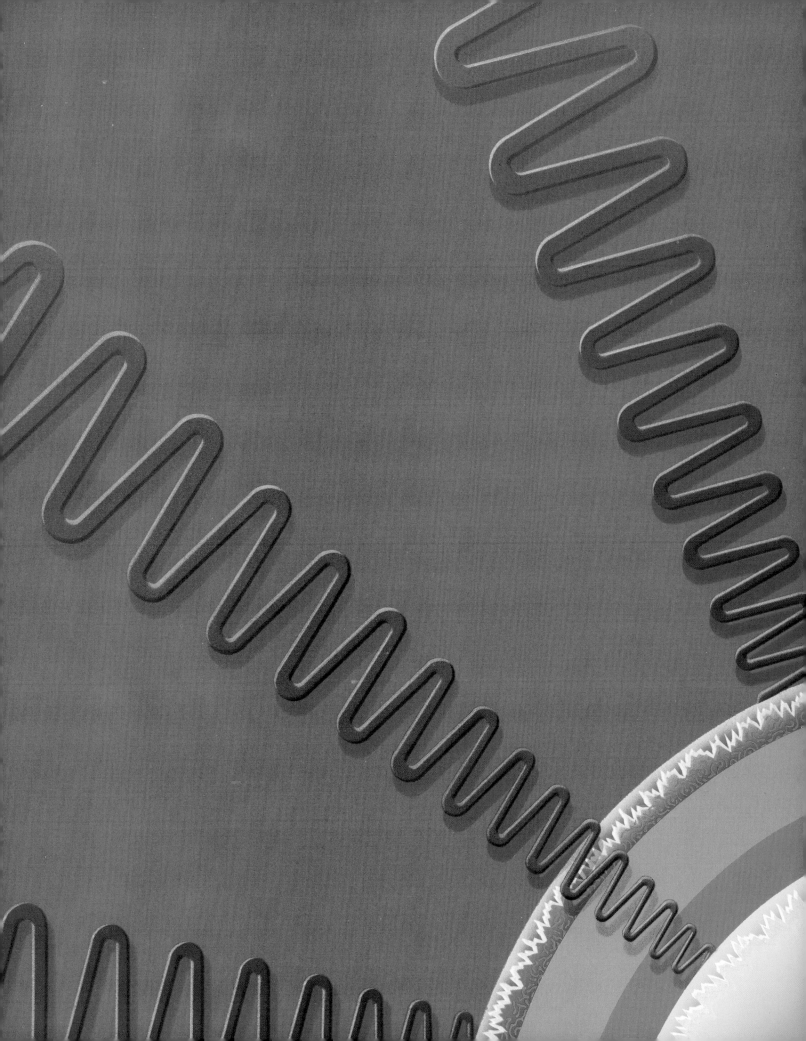

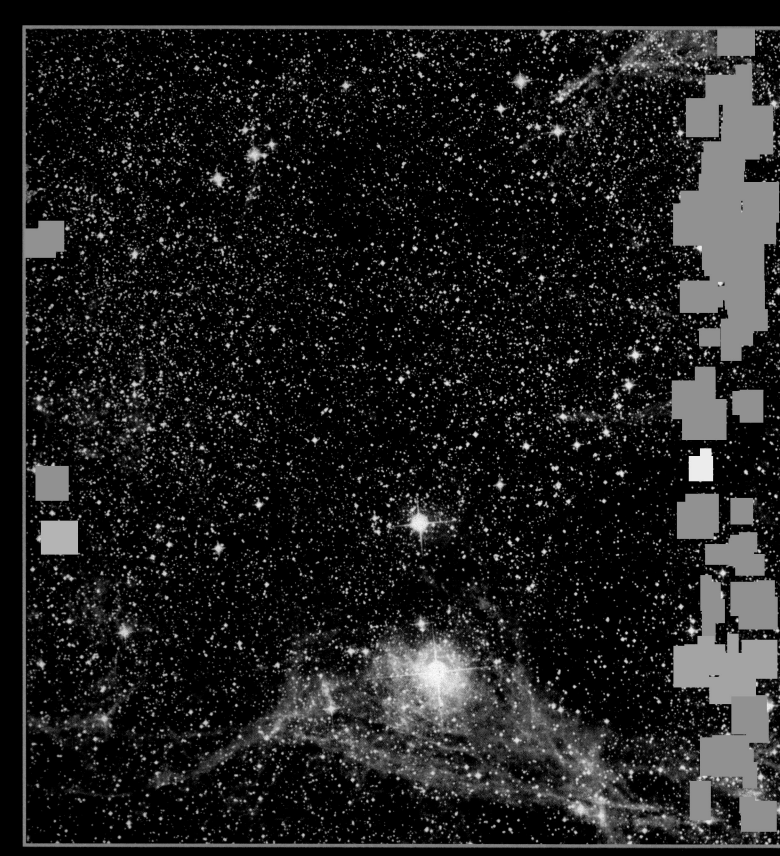

Gleaming filaments—the remains of a supernova that exploded in the constellation Vela more than 10,000 years ago—ripple through interstellar space and collide with other cosmic debris to trigger the birth of new stars.

hree o'clock on the morning of February 24, 1987, found Ian Shelton standing on a ladder in an unheated cinder-block shed atop 8,000-foot Las Campanas Mountain in desolate northern Chile. The twenty-nine-year-old Canadian was photographing the Large Magellanic Cloud, a cottony galaxy 170,000 light-years away. Close by were several high-tech white domes, one of them housing a twenty-four-inch telescope that belonged to the University of Toronto. Shelton was responsible for tending the sophisticated instrument, but because a stream of astronomers visiting the observatory complex kept the twenty-four-inch occupied, he rarely got to use it himself. So he did his stargazing in the shed, with the aid of an outmoded ten-inch telescope that had lain in disrepair for some years. Unlike more advanced motor-driven models, the ten-inch had to be adjusted by hand, minute by minute, to compensate for the Earth's rotation.

Shelton, who had twice dropped out of graduate school, lived for moonless nights like this one. As a youth in Winnipeg, he had built an observatory in his parents' backyard, and he much preferred life on the mountaintop to civilization and the constraints of higher education. Left to himself, he worked superhuman hours. Tonight, though, his tasks were not going well. He had planned to test the equipment attached to the telescope by photographing the Large Magellanic Cloud in a three-hour exposure. As he began the session, the sliding, corrugated metal roof that opened to the stars stuck shut, and he had to climb up to shove it aside. To compound his difficulties, a forty-mile-per-hour wind had been buffeting the mountain all night. Such turbulence made for poor viewing, because it blurred the images.

Suddenly, the stars began to disappear. "The wind was so loud," recalled Shelton a few months later, "I couldn't hear anything else. All these things were going through my mind: the ruin of the photographic plate, the destruction of the telescope, the end of the world." After a few moments of stumbling around in the dark, wondering whether several hours' work had been wasted, he realized what had happened. The wind had blown the recalcitrant roof shut and in the process had knocked his telescope askew. "Enough was enough," said Shelton. He decided that after developing the last plate he would call it quits and go to bed.

The first thing he saw on the plate in the darkroom was a bright spot next to a Magellanic feature known as 30 Doradus, or the Tarantula nebula. Shelton thought his plate must be marred. For twenty minutes, he tried to explain away the spot. Then he went outside. There, blazing in the darkness, was the brightest, closest supernova seen since 1604, when the German astronomer Johannes Kepler spied one in the constellation Ophiuchus (Holder of the Serpent). Shelton walked to one of the nearby domes and calmly conveyed the news to two colleagues who were still at work.

As it happened, Shelton was not the only one to have seen the supernova. Around two o'clock that morning, a Las Campanas technician named Oscar Duhalde had noticed that the region around 30 Doradus was uncommonly bright, but he had not bothered to investigate further. Other astronomers around the world were less blasé when they, too, spotted the stellar wonder, and because Las Campanas is so remote, Shelton almost failed to receive the credit for his discovery: Emissaries from the observatory had to drive to the coastal town of La Serena to get a message through to the Central Bureau for Astronomical Telegrams, the international clearinghouse for news of unusual astral events.

The telex giving details of the find arrived in the bureau's Cambridge, Massachusetts, office just barely ahead of a phone call reporting that Albert Jones, an amateur astronomer in New Zealand, had also seen the brilliant object. Furthermore, it became clear over the next few days that another astronomer, Robert McNaught at Siding Spring Observatory in Australia, had actually been the first to capture the supernova on film, when its brightness was magnitude 6, as faint as can be seen with the naked eye. Unfortunately, McNaught had failed to develop his plate the same night, and the trio of Shelton, Duhalde, and Jones will go down in astronomical history as codiscoverers of a major supernova.

It had been a long dry spell. Evidence from other galaxies indicates that a star goes supernova perhaps once or twice every century in the Milky Way, but clouds of dust and gases strewn throughout the galaxy keep their light from reaching the Earth. In the years since Kepler saw the last exploding star, astronomers have found none in the Milky Way. Of course, SN1987A occurred in a neighboring galaxy—a diminutive satellite of the Milky Way—but the supernova was close enough and bright enough to send astronomers throughout the Southern Hemisphere scurrying for the nearest observatory to study the "guest star," as ancient Chinese court astronomers referred to such phenomena. Said one scientist, "It's so exciting, it's hard to sleep."

Modern astronomy has formulated and discarded dozens of theories about how stars live and die. Yet this was the first real opportunity to test some major ideas about stellar death. Theorists had determined, for example, that only those stars beginning life with roughly ten times the mass of the Sun (ten solar masses) could end as supernovae. Pinpointing the progenitor of SN1987A, they were sure, would help verify this notion. Furthermore, spectral analyses of radiation from the blast might help confirm theoretical models charting the late stages of nuclear burning in stellar furnaces and trace the steps leading to a star's cataclysmic explosion.

BEYOND THE MAIN SEQUENCE

The discovery and subsequent analysis of SN1987A provided a spectacular climax to a series of astronomical observations and interpretations that had started about forty years earlier. At the beginning of that period, astronomers were agreed on the overall meaning of the Hertzsprung-Russell diagram. They knew that aging stars left the main sequence to become red giants and then white dwarfs (unless their core masses were above the Chandrasekhar limit of 1.44 solar masses, in which case they collapsed from red giants into some then-unknown object). But the details of that crucial transition from the main sequence, when stars exhausted their stellar fuel, were still a mystery. In the years immediately following World War II, astronomers turned their thoughts to this question. One of the first to attack the problem was Martin Schwarzschild, a man with a self-professed passion for detail work.

Schwarzschild, the son of respected German astronomer Karl Schwarzschild, received his doctorate from the University of Göttingen in 1935. The following year, the oppressions of the Nazi regime drove him from his homeland to Norway, where he held a research post for a year before emigrating to the United States. Schwarzschild eventually became a U.S. citizen and spent most of his diverse and highly productive career at Princeton University. His central interest was how a star's structure evolves over time.

Schwarzschild was one of the first astronomers to use computers. His major work on late stellar evolution, which began in 1948, employed primitive machines, including a notoriously breakdown-prone device built by Princeton colleague John von Neumann, a pioneer in computer design. The astronomer drew on existing knowledge to come up with four complex equations describing the relationships among stellar mass, luminosity, temperature, and chemical composition. Von Neumann's machine then worked out how the solutions for those equations would change over the long life span of a star. The intricate mathematical modeling revealed that stars in the late stages of their development were very different beasts from their stable, main sequence incarnations.

First Schwarzschild established the process by which a star of any size becomes a red giant. During a star's years on the main sequence, it burns its hydrogen fuel steadily and changes only minimally in brightness and temperature. All the while, it balances two countervailing forces: the gravity that tugs it constantly toward collapse and the outward pressure created by thermonuclear conversion of its own gas. Buoyed mainly by the heat created when hydrogen fuses into helium, a main sequence star can win the battle and

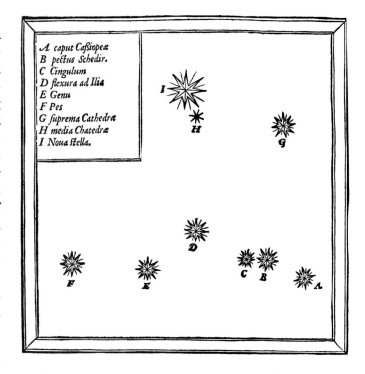

To record rare stellar explosions for posterity, sky watchers have used whatever techniques were at hand. The image above, for example, is from a woodcut by the Dane Tycho Brahe, who carefully plotted the location of the 1572 supernova (marked "I") near the constellation Cassiopeia. At right is a collage of images that span nearly a thousand years: The bottom portion is a pictograph from the American Southwest that may represent the supernova of 1054, which spawned the Crab nebula. A modern astronomer used twentieth-century special effects to juxtapose the pictograph with a photographic re-creation of the event that inspired it.

stay inflated, but as soon as the atomic hurly-burly at its center slows, the core begins to fall inward.

With astronomer Allan Sandage, then a graduate student at the California Institute of Technology and later a famed cosmologist, Schwarzschild charted this collapse. According to their numerical studies, a star's core begins to shrink after all its hydrogen fuel is used up. Bereft of hydrogen, but filling with extra helium from the burning of hydrogen in the shell around it, the core continues to heat up. The additional heat energy causes the envelope of unreacted hydrogen outside the core to expand. The star swells, ultimately becoming a bloated version of its former self, hundreds of times larger in diameter. Its surface cools and its light reddens. The star is now classified as a red giant or supergiant, depending on its initial mass, and occupies the so-called giant branch of the Hertzsprung-Russell diagram.

When the core temperature reaches about 100 million degrees Kelvin, a fresh thermonuclear reaction starts up: helium burning. Three nuclei of helium fuse to create a single nucleus of carbon and two particles known as gamma ray photons. During the star's helium-burning lifetime, which is only a fraction of its previous hydrogen-burning span, the stellar giant continues to swell. The Sun, for example, will eventually swallow the rocky inner planets Mercury, Venus, and maybe Earth and evaporate the gases that swathe the outer planets. It will shine a hundred times more brightly than it does today.

Once he had established the overall picture of the red giant phase, Schwarzschild turned his attention to the critical period in which the star shifts from burning hydrogen to burning helium. Joining him in 1955 was a somewhat unlikely collaborator: a maverick British astronomer and mathematician named Fred Hoyle.

Whereas Schwarzschild thrived on detail, Hoyle's forte was conceptualization. From youth, he shared with his north Yorkshire compatriots a persevering individualism and disregard for the English class system. During his grammar-school years, he largely educated himself. Often he could be found truant in his mother's kitchen, performing complicated and sometimes dangerous experiments out of a chemistry text that his father, a cloth merchant, had purchased for his own edification. Eventually he won two science scholarships to Cambridge University. There Hoyle associated with some of the finest scientific minds of his time, including the formidable theoretical phys-

icist Paul Dirac, and went on to win a faculty post at St. John's College. All
the while, he sowed his ideas, nurtured by a prolific and sometimes unorthodox intellect, throughout the astronomical community.

Hoyle's most productive alliances came with those who, like Schwarzschild,
possessed the mathematical expertise to balance his imagination. He and
Schwarzschild determined that there was a key difference between stars
below about two and a half solar masses and stars above that critical limit.
The higher-mass stars move fairly smoothly from hydrogen burning to helium
burning. But their lower-mass cousins are not so tranquil. As Hoyle and
Schwarzschild studied globular clusters, spherical groupings of hundreds of
thousands of older stars that dot the outskirts of the Milky Way, they theorized that the hydrogen-to-helium transition inside lower-mass stars occurs
with a series of conflagrations.

According to Hoyle and Schwarzschild's calculations, as pressure increases
in a red giant's core, electrons are stripped from atoms, and over millions of
years gravity packs these free particles so close together that they cannot
be compressed further. (In this state, a gas is said to be degenerate and does
not behave as it usually would.) When the interior temperature of the dying
star reaches 100 million degrees Kelvin, helium nuclei begin to fuse into
carbon, releasing heat, which causes more helium nuclei to react. But the
compressed free electrons do not react as quickly to the heat, so instead of
expanding, as would normally happen, the core remains the same size even
while its temperature doubles or triples. The rate of fusion climbs as the
energy stays trapped within the core. At its peak comes a burst of fusion—
a "helium flash"—and an accompanying flareup of heat that unleashes the

By examining the spectra of stars in
so-called globular clusters like M13 *(top)*,
Martin Schwarzschild developed theoretical models to explain several aspects of
stellar evolution, including the phenomenon known as the helium flash. Because
stars in globular clusters are much older
and packed a thousand times closer
together than those in the solar neighborhood, astronomers can gather a variety
of information about many stars formed
early in the life of the galaxy.

packed electrons, causing the core to expand very quickly and releasing energy equal to the output of an entire galaxy.

The flash has the immediate effect of catapulting the degenerate atoms back into a normal state. But over time, as gravity again takes hold, the core matter may again become degenerate. As the outer layers of the star gradually cool and lose mass to stellar winds, additional, though rather less energetic, flashes may come with greater frequency as conversion of helium to carbon pushes the core temperature still higher. At this point the star leaves the realm of red giants in the Hertzsprung-Russell diagram to move back toward, but not onto, the main sequence.

Astronomers have no doubt that this picture of helium flashes is right, but they cannot hope ever to see the flareups. Although scientists are able to trace their consequences by measuring a star's energy output, the helium flashes are forever shielded from view by millions of miles of opaque gases.

DEATHS OF RED GIANTS

In the early 1960s, Hoyle turned his attention to other enterprises, including the writing of a series of popular science-fiction novels. Schwarzschild, for his part, went on, with Richard Härm of Princeton, to examine the further evolution of a low-mass red giant—the point at which it becomes a white dwarf star. Schwarzschild found that when helium conversion slows, after about 100 million years, the ebbing of the fire at the star's center ushers in a period of instability. Carbon and oxygen, the by-products of helium fusion, clot the core, and once again the star loses the ability to support itself; gravity is ascendant. As before, the core begins to collapse. The collapse in turn creates enough heat to ignite helium in a surrounding shell that encloses a volume somewhat smaller than that of the Earth.

In a star with up to two and a half times the mass of the Sun, the outer layers—driven by the renewed combustion of helium—balloon until the star is about 240,000,000 miles in diameter (roughly the diameter of the orbit of Mars). It has become a red giant for a second time, moving back to the so-called giant branch of the Hertzsprung-Russell diagram. At this point, the star is very close to death.

Mathematically modeling the next steps in the stellar evolution proved extraordinarily difficult. The calculations indicated that another sort of helium flash would start taking place, this time at a distance from the center, whenever the temperature rose slightly in the helium shell surrounding the star's core. For brief periods, at intervals of roughly 100,000 years, the star's energy output would rocket a thousandfold owing to a series of such shell flashes, which astronomers now call thermal pulses. Eventually, the pulses might help to tear the shell away from the core. The outermost layers, containing as much as 80 percent of the star's mass, would be propelled into space in a diffuse cloud. The hot core would then light up the last remnants of gas, resulting in a glowing cloud around a blue-white star. To eighteenth-century astronomers, whose telescopes lacked the power to resolve the images, the visible interiors of these clouds resembled the disks of planets. Hence they were dubbed planetary nebulae. Modern astronomers have cataloged more than 1,000 planetaries and estimate that there are some 50,000 in the Milky Way galaxy alone, most of them with irregular shapes.

This, then, was the ultimate fate of roughly Sun-size stars. They would shed most of their material into the interstellar medium and spend the rest of their days cooling and shrinking until they were only a few thousand miles across, ending as white dwarfs. A few white dwarfs, however, show blazing signs of their existence before they finally wink out. Those in binary systems may flare brilliantly for a few days, increasing their luminosity by factors ranging from hundreds to millions, before fading back to normal within a few months to a year. Scientists believe the stars constitute a type of stellar flambé: Matter falling onto them from companion stars provides the fuel for the flareups. Each event of this type appears like the sudden formation of a new star; hence, they are called novae, from the Latin word for "new." Astronomers typically spot two to three novae per year.

Even these stars eventually exhaust their fuel and cool down. Recent calculations indicate that white dwarfs undergo one final transformation before they are truly dead: The almost motionless atomic nuclei in the core succumb to electrical forces that bind them into rigid lattices, releasing a brief burst of energy as the star crystallizes. Then the star cools more rapidly, growing faint and finally flickering into darkness. Astronomers theorize that the process takes many billions of years, so although they are keeping an eye out for such objects in the Milky Way, the galaxy may not be old enough for very many of its mid-size stars to have reached this stage. Moreover, stars with only a fraction of the Sun's mass have such low gravity and such slow rates of converting hydrogen into helium that they may take as long as 20 trillion years to traverse the path to white dwarfdom.

In any event, not all stars share this fate. Even while Schwarzschild was tracing the progress of smallish stars, some of his colleagues were demonstrating that more massive objects are very different creatures. Large stars, they would find, run through their nuclear fuel at a furious clip and die young in monumental explosions, or supernovae.

WILD IDEAS

The notion of exploding stars seemed so outlandish when first proposed that most astronomers rejected it. Recognition of its merit came only after a major effort by a dogged researcher whose behavior was at least as eccentric as the idea itself. Born in Bulgaria and raised in Switzerland, Fritz Zwicky came to the United States in 1927 and settled into a faculty post at the California Institute of Technology in Pasadena. Trained in mathematics and physics, he exercised a wide range of scientific interests, addressing technological problems as well as theoretical ones. He was a space buff who boasted that he had sent the first human artifact beyond Earth's orbit: Early in the space age he had put an explosive charge atop an Aerobee rocket. When the rocket reached

Fritz Zwicky, who in 1931 coined the term supernova during a lecture at the California Institute of Technology, was one of the first scientists to undertake a systematic search for these explosive outpourings of stellar energy. Over a three-year period from 1936 to 1939, Zwicky photographed hundreds of galaxies and discovered twelve supernovae in the process.

the peak of its trajectory, he triggered the charge to fire a scrap of metal off into deep space. This enterprising experimentalist eventually held about fifty patents, including one for an underwater ramjet he called the hydrobomb.

Many of Zwicky's fellow astronomers feared him, because he believed himself superior to all of them both physically and mentally. He frequently tried to demonstrate the former by doing one-handed push-ups on the floor of a Caltech dining room, challenging all comers to emulate him. He showed his disdain for his colleagues' mental capacities by referring to them as spherical bastards—"because they are bastards every way I look at them." Walter Baade, an eminent German-born astronomer at the Mount Wilson Observatory north of Pasadena and one of his collaborators, was said to believe that the eccentric scientist intended to murder him.

It was hardly surprising, then, that Zwicky had difficulty persuading the astronomical establishment to accept his ideas. The fight began quietly during 1931 in a class he taught on novae at Caltech. For decades astronomers had been noting that certain novae were much brighter than the average. Zwicky decided that these represented an entirely new breed, which he called "super-novae" (later dropping the hyphen). These stars were more luminous and flared for nearly a month rather than the few days typical of a nova. Zwicky also suspected that the origins of supernovae were different from those of novae, although he did not know at the time that novae are associated exclusively with white dwarfs.

He went public with the distinction in 1933 at a meeting of the American Physical Society. With coauthor Walter Baade, he announced in a paper that "supernovae flare up in every galaxy once in several centuries. The lifetime of a supernova is about 20 days and its brightness at maximum may be as high as 100 million times that of the sun. . . . In the supernova process, mass in bulk is annihilated."

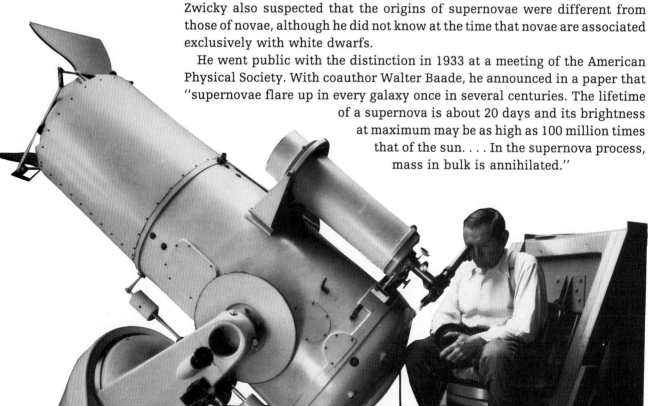

The response was universally skeptical. By what mechanics was the precipitous collapse of huge stars turned into a gargantuan explosion? And if the phenomenon was so common, why had astronomers over the centuries observed only a handful of events that met the criteria defined by Zwicky? Undaunted, Zwicky set out to find examples. As he remembered it later, he was accompanied "by the hilarious laughter of most professional astronomers and my colleagues at Caltech." From the rooftop of a building on campus, he aimed his camera at a star-thronged region near the constellation Virgo. When this effort failed to turn up anything after two years, he persuaded the director of the observatory on Mount Wilson to let him use a new type of telescope that would enable him to photograph larger chunks of the sky.

With this instrument, he and a coworker recorded a startling burst four million light-years away. "Beating the tar out of the sky," Zwicky noted afterward, "I found my first supernova in the spiral galaxy NGC 4157, in March 1937." Five months later, on August 26, he spotted a second explosion elsewhere. During the next three years, he took 1,625 photos of 175 regions of the sky and discovered twelve supernovae; eventually his personal score reached 120. Between 1958 and 1975—a year after Zwicky's death—a formal continuation of these first searches was carried on with a forty-eight-inch telescope at Palomar Observatory. The final tally was 281 events, including those discovered by Zwicky himself. To date, astronomers around the world have seen some 600 of the exploded stars.

Zwicky had succeeded in establishing that supernovae were not the impossibilities his colleagues had claimed. When stars begin life with anywhere from about ten to a hundred solar masses (the size of the largest stars yet seen), they seem to come to a violent end; in a typical galaxy this would happen an average of once every fifty years. Since Zwicky's pioneering searches, astronomers have classified supernovae into two types. Type I events, seen mainly in ancient galaxies, show a rapid rise in luminosity that then fades slowly and steadily over several months. This type is believed to result from thermonuclear explosions in binary pairs that completely disrupt the evolved cores of one or both of the stars. Type II events, which occur in parts of galaxies inhabited mainly by young stars, seem to represent the obliteration of massive stars that, after running through a series of nuclear reactions, have reached the unimaginably high temperatures of more than three billion degrees Kelvin in their cores.

STELLAR INTERIORS

It had been a long road from the definition of white dwarfs in the 1920s to the discovery of supernovae in the late 1930s and 1940s. By midcentury, scientists' general understanding of stellar history was almost complete. But British astronomer Fred Hoyle had some unfinished business. In 1948, George Gamow, principal author of the Big Bang theory, had proposed that the fireball that created the universe also forged hydrogen, helium, and heavier elements. Hoyle, however, was a proponent of the competing Steady State

theory, which asserted that there was no initial explosion: Instead, matter is continually appearing from nothing throughout the universe. Hoyle was thus intent on finding another mechanism for the synthesis of most elements. He decided that substances heavier than helium had to have been made in a step-by-step fashion within the fast-burning furnaces of gargantuan stars.

A young Caltech professor hearkened to Hoyle's hypothesis. William Fowler, a cherubic, gregarious midwesterner with a passion for steam locomotives and Pittsburgh Pirates baseball, became convinced that "the grand concept of nucleosynthesis"—the creation of progressively heavier elements within aging stars—was correct. As a freshman at Ohio State University, Fowler ("Willy" to his friends) intended to become a ceramics engineer but was sidetracked by physics. He went on to graduate school at Caltech and after World War II devoted himself to the intricacies of nuclear astrophysics. "Edwin Salpeter of Cornell University came to visit in the summer of '51," Fowler recalled more than thirty years later, "and showed that the fusion of three helium nuclei to create one carbon nucleus could probably occur in red giants, but not in the Big Bang." Further experiments by researchers in Pasadena confirmed that the temperatures and densities inside red giants were indeed sufficient to the task. Fowler promptly undertook a project with Hoyle that aimed to account mathematically for the varied chemical compositions of stars, as manifested in their spectra.

Shortly thereafter, in 1954, he traveled to Cambridge, where he and Hoyle spent the year amplifying the calculations. Joining them were a British husband-wife team whom colleagues affectionately called "B squared" (B^2), Geoffrey Burbidge and Margaret Burbidge. Geoffrey, who had begun his career in theoretical physics, was lured to heavenly pursuits by Margaret when he was a graduate student and she an undergraduate at the University of London. The daughter of a chemist, Margaret had an incisive intellect and a remarkable aptitude for observational research. Despite this, she was forced to struggle for much of her career to win equal footing in the male-dominated world of astronomy. When the couple first moved to the United States in 1955, for example, Margaret could only get viewing time at the Mount Wilson telescope by pretending to be her husband's assistant.

When Fowler returned to the United States in 1956, his British cohorts joined him, and the inquiry continued full bore at Caltech. The following year, the quartet's seminal paper "Synthesis of the Elements in Stars" was published. In astronomical circles, it became so famous that it was assigned the shorthand designation "B^2FH" for the initials of the four authors' surnames. Although it did not confirm Hoyle's beloved Steady State hypothesis, this classic study did chronicle how nuclear processes could yield heavier and heavier elements from carbon all the way to uranium.

Fifteen billion years ago, the theory went, the universe was a vast, expanding cloud of hydrogen. In this great fog, gravitational perturbations arose and stars coalesced, made almost entirely of hydrogen. Deep in their interiors, protons and electrons collided to form deuterons, the nuclei of

In 1957 four researchers working at the California Institute of Technology published a paper that offered the first detailed analysis of nucleosynthesis, the stellar process that produces elements heavier than hydrogen. Collectively referred to by their colleagues as B^2FH, the four were (top to bottom) William Fowler, Fred Hoyle, Margaret Burbidge, and Geoffrey Burbidge.

MESSAGES FROM SN1987A

When a supernova was discovered blooming at the edge of the Large Magellanic Cloud early in the morning of February 24, 1987, the astronomical world leaped into action. Within hours, nearly all the radio and optical telescopes in the Southern Hemisphere were trained on SN1987A, as it came to be known. Within a day, the *International Ultraviolet Explorer* satellite was recording radiation from the blast. Within weeks, the Japanese x-ray satellite *Ginga* began observing the scene, followed by instruments aboard NASA's *Solar Maximum Mission* satellite and the Soviet *Mir* space station. Soon scientists knew more about this one supernova than they did about all previous stellar explosions.

Some findings confirmed long-standing theories—that elements heavier than hydrogen are produced in a supernova explosion, for instance, and that the event also releases neutrinos *(below)*. Others were confounding, at least at first. For example, the visible light emitted by the star in its first year peaked at an intensity ten times dimmer than expected *(page 84)*. This observation may be explained by a greater surprise: the identification of a blue supergiant star, Sanduleak −69° 202, as the supernova's progenitor. Theorists who had believed that only red supergiants experienced core collapse in so-called Type II supernova explosions had to revise their thinking in light of all the new information that was streaming toward the Earth. A sampling of that data, gathered by a variety of instruments, is shown on the following pages.

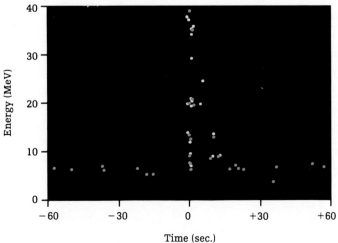

 Neutrinos. The first sign of a supernova appeared some eighteen hours before observers saw SN1987A in the sky: Two underground water tanks—one near Cleveland, Ohio, and the other in Kamioka, Japan—registered the arrival of a few tiny particles called neutrinos, generated in the explosion. They went unnoticed until scientists learned of the supernova and checked the tanks' records. Physicists estimate that the supernova sent billions of neutrinos streaming through every square inch of the Earth, but only a handful announced themselves by striking water molecules in the two locations. Each collision produced a flash of light detected by photomultiplier tubes lining the tank walls. The graph at left shows the marked increase in collisions recorded simultaneously in Ohio *(yellow)* and Kamioka *(blue)*. Readings below seven million electron volts (MeV) in Kamioka and below twenty MeV in Ohio are background radiation.

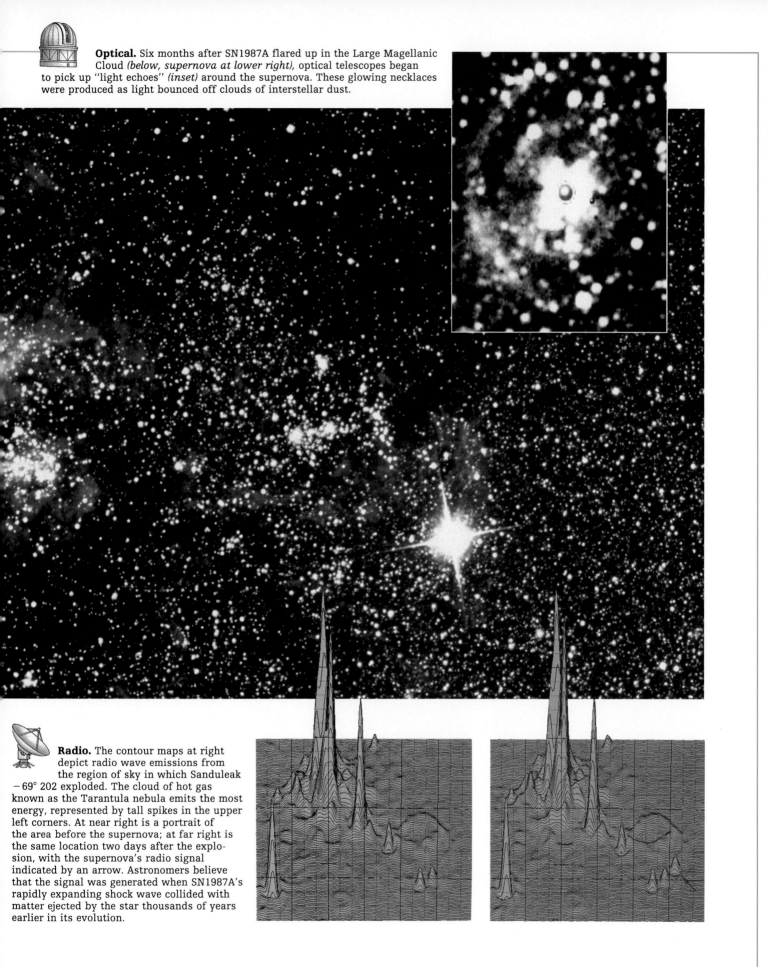

Optical. Six months after SN1987A flared up in the Large Magellanic Cloud *(below, supernova at lower right)*, optical telescopes began to pick up "light echoes" *(inset)* around the supernova. These glowing necklaces were produced as light bounced off clouds of interstellar dust.

Radio. The contour maps at right depict radio wave emissions from the region of sky in which Sanduleak −69° 202 exploded. The cloud of hot gas known as the Tarantula nebula emits the most energy, represented by tall spikes in the upper left corners. At near right is a portrait of the area before the supernova; at far right is the same location two days after the explosion, with the supernova's radio signal indicated by an arrow. Astronomers believe that the signal was generated when SN1987A's rapidly expanding shock wave collided with matter ejected by the star thousands of years earlier in its evolution.

Wavelength

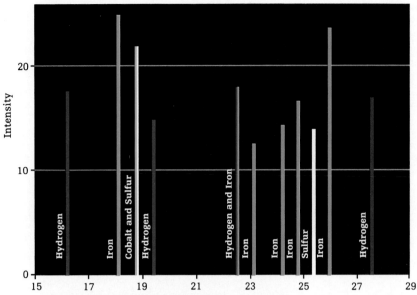

Infrared. Observations in the infra-red region of the spectrum, some recorded by a telescope aboard a C-141 jet transport, yielded crucial information on elements formed in the supernova explosion. Initial readings detected hydrogen emissions from the star's outer layer. Nine months after the supernova appeared, scientists identified radiation from heavy elements such as iron and sulfur *(above)* that were created as the star's core collapsed and exploded.

Gamma rays. Gamma radiation from the decay of radioactive isotopes of cobalt, first detected by the *Solar Maximum Mission* satellite, is revealed by an image from an instrument aboard a high-altitude balloon. The finding indicates that new isotopes are created during the explosion. With a half-life of seventy-eight days, the cobalt would have decayed into iron by the time of the explosion if it had come into being earlier.

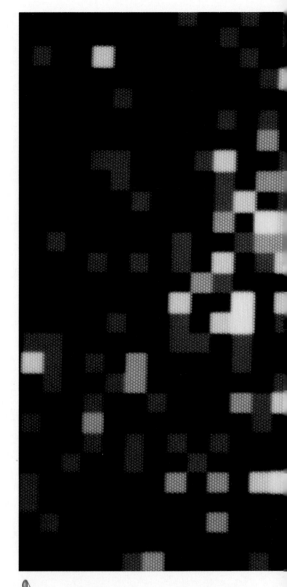

Optical. Digitized readings *(above)* taken by optical equipment aboard the *International Ultraviolet Explorer* satellite show SN1987A's brightest areas in red, with progressively dimmer regions in green, blue, white, gray, and black. Astronomers used this data to plot the supernova's light intensity over the course of a year *(graph, below)*.

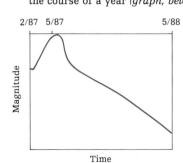

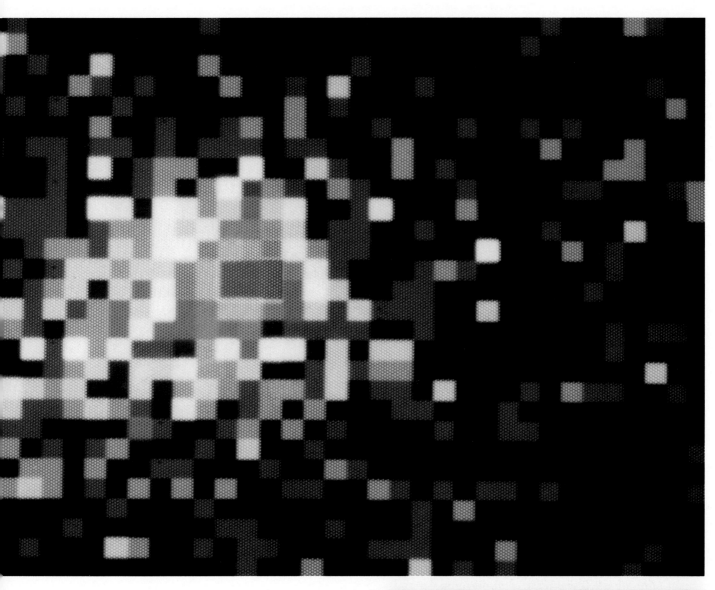

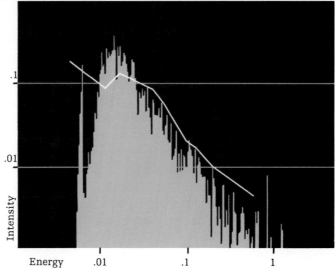

X-ray. A graph of x-ray emissions from the supernova *(right)* shows a remarkable congruence between a complex theoretical model (the series of jagged peaks) and observations from equipment aboard the Soviet Union's *Mir* space station (the white line). This portrait of the explosion's x-ray spectrum shows the intensity of emissions (vertical axis of graph) as well as their energy in millions of electron volts (horizontal axis). The findings closely matched scientists' understanding of how matter from a star's interior mixes with its outer layers, allowing gamma rays from the decay of radioactive elements to interact with the surrounding matter and then degrade into x-rays.

hydrogen isotopes, which in turn collided, eventually to form helium. As the stars aged and their cores heated up, a succession of heavier elements were synthesized in their centers: carbon, nitrogen, oxygen, neon, and other substances up to iron. According to B[2]FH, "The temperature is not everywhere the same inside a star, so that the nuclear evolution is most advanced in the central regions and least or not at all advanced near the surface. Thus the composition of the star cannot be expected to be uniform throughout."

Some of the heavy elements flowed out into space as older stars lost mass through stellar winds. But the biggest members of this first generation of stars, said the researchers, burned furiously and exploded as supernovae, spewing out their heavy elements. The ejected material was recycled—incorporated in a second and eventually a third generation of stars, including the Sun. Fusion in these younger, metal-enriched stars proceeds a little differently, allowing them to manufacture still other elements, including such rarities as californium and gold. At the end of its life, a massive star of this generation is layered like an onion, with a thick hydrogen sheath encasing successively thinner layers of helium, carbon, oxygen, and silicon, all of which are peppered with magnesium, calcium, and other elements. At the center is an iron core no more than a few thousand miles in diameter and so stable that nuclear reactions have ceased. The core's density at this point is some 20,000 to 200,000 tons per cubic inch, and the pull of gravity is immense.

Even at this stage, the core is growing in mass, as silicon burning in the shell immediately outside it creates additional iron. Eventually, the core's mass reaches the Chandrasekhar limit, setting off the star's final death throes. The inner part of the core implodes within one one-hundredth of a second, collapsing to a tremendously dense sphere and then rebounding at high speed to collide with the still-collapsing outer portion of the core. The shock wave generated in this collision rips the star apart, throwing out high-energy radiation and elementary particles known as neutrinos. The colliding gases fuse to form even more elements, such as isotopes of nickel and cobalt. Once again, a supernova flings its treasury of heavy elements into the interstellar reaches to form the seeds of complex molecules, planets—and life. In a very real sense, one astrophysicist has said, "We are the grandchildren of supernovas."

THE VALUE OF SN1987A

The nucleosynthesis theories of Hoyle, Fowler, and the Burbidges became widely accepted in the thirty years after their publication, with the true stamp of approval coming in 1983, when Fowler was awarded a Nobel prize for his leading role in the work. Still, these and other theories concerning stellar death remained largely conjectural. Astronomers could study the remnants of several supernovae, but to test their theories in detail they needed a star in the earliest stages of going supernova—which explains why sky-gazing scientists reacted with such excitement to the advent of SN1987A. They hurried to deploy an arsenal of detectors, from high-resolution, supercooled

Canadian astronomer Ian Shelton, shown here at the eyepiece of a telescope at Las Campanas Observatory in northern Chile, captured one of the earliest images of SN1987A, the nearest supernova to appear for almost 400 years. A faint streak just above the telescope signals a far more common celestial phenomenon: a passing meteor.

spectrometers for detecting gamma rays to so-called charge-coupled devices, integrated circuits that function as highly sensitive cameras. In the months following the blast, scientists observed the stellar remains with equipment on high-altitude balloons launched from the arid Australian outback. They employed instruments aboard the orbiting *International Ultraviolet Explorer* and *Solar Maximum Mission* satellites. Observatories in Chile, New Zealand, Australia, South Africa, and elsewhere looked at emissions in both the visible and infrared wavelengths, which they had been unable to do in the past with more distant supernovae. Physicists mapped energy outputs and strove to comprehend behaviors that did not match their theoretical expectations.

As soon as possible, astronomers examined star maps to determine which star was responsible for the disturbance. The culprit turned out to be Sanduleak $-69°$ 202, a star of about twenty solar masses and intensely blue in color. Here was an oddity: Supernovae were thought to arise from red supergiants, and Sanduleak had been a blue supergiant, a well-evolved star with a much smaller diameter than those of the red variety. In another anomaly, the pattern of radiation from SN1987A did not fit the standard curve of either a Type I or a Type II event. By monitoring more distant supernovae, astronomers had established that most reach their maximum brightness within a short time of exploding, leaping three or four magnitudes in a few days. SN1987A, though, became dimmer and cooler for the first seven days before brightening and warming. It did not attain its peak until eighty-five days after showing up on Ian Shelton's photographic plate. Then, strangely, it assumed the appearance of a normal, Type II explosion.

Sorting through the anomalies, astronomers realized that accepting the possibility that a blue supergiant could go supernova—and one obviously had—gave them a ready explanation for the peculiar pattern of emission. In a sense, a blue supergiant is simply a more compact version of a red supergiant: The physics of the two cores are essentially the same. However, being denser, Sanduleak $-69°$ 202 was less easily torn apart by the shock wave traveling outward from the collapsed core. Much of the energy that would have dispersed into space was spent in the basic demolition. Thus the energy output rose more slowly than anticipated.

About six months after the first flare, the *International Ultraviolet Explorer* picked up emissions from a haze of matter approximately one light-year away from

Sanduleak $-69°$ 202's last known position. Some astronomers believed that this matter, which was rich in nitrogen, had been ejected from the star some 20,000 years before its final explosion—evidence of a previous incarnation as a red supergiant.

In December 1987, almost a year after the supernova's discovery, *Solar Max* picked up gamma rays coming from the Large Magellanic Cloud. On the ground, there was elation: The B^2FH argument about nucleosynthesis predicted that gamma radiation, produced by the decay of radioactive isotopes created in the instant of the supernova burst, would accompany a supernova flareup. After thirty years, champions of the nucleosynthesis idea had the final confirmation of their theories.

By far the most gratifying aspect of supernova 1987A, however, was the capture of elusive neutrinos from the event. In the late 1960s, theorists had predicted that neutrinos would escape from an exploding star over a period of several seconds as its iron core imploded, carrying away roughly 99 percent of the explosion's energy. Neutrinos are tantalizingly difficult to detect. They have little or no mass, they possess no electric charge, and they can penetrate a solid object like the Earth as if it were not there. Undaunted, physicists around the world have built several devices designed to register the ghostly passage of neutrinos. Their main goal has been to determine whether protons, the stablest of elementary particles, decay as some theorists contend they do. The detectors, giant pools of purified water deep underground, are designed to spot the components of decomposing protons—which, according to theory, would include neutrinos. It occurred to researchers that the detectors might also have a second use: If a supernova ever appeared sufficiently nearby, they could pick up neutrinos originating in the blast.

Sure enough, two detectors, Kamiokande-II, installed in a former mine in Japan, and IMB, near the southern shore of Lake Erie in Ohio, began receiving signals at 7:35:35 universal time, or 2:35:35 p.m. eastern standard time, on February 23 hours before the supernova became visible. At Kamiokande-II, eleven neutrinos arrived in the space of 12.5 seconds; eight came in 5.5 seconds at IMB, which is named for its three sponsors, the University of California at Irvine, the University of Michigan, and Brookhaven National Laboratory. It was astonishing enough that the particles had hit the two tanks almost simultaneously after a 170,000-year journey. What they told astronomers about the energy released by SN1987A was even more breathtaking. In a single second, the explosion had put out a hundred times more energy than the Sun will have emitted in its entire 10-billion-year lifetime.

The informational bounty of SN1987A is not yet exhausted. In a few years, when the initial radiation and debris have cleared away from the area, theorists hope to detect the abandoned core of the supernova. Astronomers will train various kinds of telescopes on the scene of the explosion in search of the weird, tiny, fantastically dense object known as a neutron star. If they find it, the supernova will have provided yet another confirmation of the strangeness of the cosmos.

COUNTDOWN TO CATASTROPHE

One moment the star was an undistinguished speck in the Southern Hemisphere skies *(above, left)*. Then it flashed into history as supernova 1987A *(above, right)*. Occurring only 170,000 light-years away in the Large Magellanic Cloud, the stellar explosion was the nearest in almost four centuries and a once-in-a-lifetime chance for astronomers to refine their theories about such paroxysmic star death. From the study of more than 600 supernovae, virtually all of them in galaxies much more distant, scientists have identified two main varieties, illustrated on the following pages. SN1987A is an example of the slightly more common Type II explosion, which takes place when the process known as core collapse liberates enough energy to blow apart the outer layers of a single massive star. A Type I supernova is thought to occur in a binary system: When matter flowing from one star onto the other reaches a critical mass, it sets off a thermonuclear blast that destroys one or both partners.

The radiation and matter flung into space by either type of supernova yield valuable clues about the cosmic recycling system, a system integral to life itself: From the death throes of the most massive stars in the universe comes material that makes up not just future stars but also planets, moons, and an infinite variety of living things.

Hydrogen Envelope

Helium
Core

1 End of the main sequence. As the last hydrogen atoms in the core fuse into helium, after a brief stellar lifetime of 9 or 10 million years, nuclear reactions temporarily cease in the center of the star. The core's resistance to the force of gravity lessens slightly, and its atoms are squeezed closer together. The core heats up from its longstanding temperature of 40 million degrees Kelvin to 170 million degrees. As this tremendous heat radiates outward to the hydrogen envelope, the envelope expands to roughly 100 times its original diameter. At its surface, the star cools, changing color from blue-white to red. Stellar winds begin to tear away several solar masses of hydrogen from the outer layer.

The End of a Superstar

The star that exploded to become SN1987A was called Sanduleak −69° 202, known by its listing in a catalog compiled by astronomer Nicholas Sanduleak. It spent most of its existence as a blue main sequence star, measuring about twenty times the mass of the Sun. Like a candle burning at both ends, a star of such enormousness fuses its supply of hydrogen at a spectacular rate—some 20 trillion tons of hydrogen atoms per second—and dies at the tender age of about 10 million years. Sanduleak −69° 202 was just such a babe when it went supernova; by contrast, the Sun, an average-size star, is already 500 times older and is only halfway through its projected life span of 10 billion years.

As long as a huge star's fuel lasts, the fusion process creates enough pressure within the core to counter the inward gravitational pull exerted by its great mass. But when the store of hydrogen at the core is used up, the balancing act goes wildly awry. Externally, the star swells into a red supergiant *(left)*. Internally, the core yields to gravity and begins contracting under its own mass, growing hotter and denser as its atoms are crushed together.

A fresh series of nuclear reactions then begin *(pages 92-95)*, each new round temporarily halting the complete collapse of the core. The first round takes a million years, but the pace quickens drastically. Fusion's final effort against gravity lasts only a few days, and the ultimate steps in the cataclysm occur in mere fractions of a second.

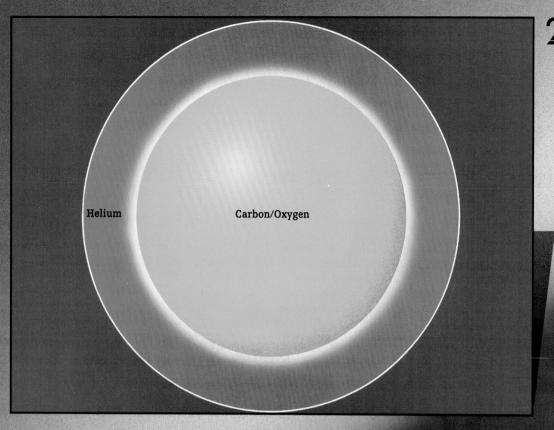

Helium

Carbon/Oxygen

2 **One million years to go.** From this point on, almost all important reactions occur within the star's core. When gravitational collapse in the core pushes its temperature past 170 million degrees, a new series of fusion reactions begin. Helium atoms fuse to form heavier elements, primarily carbon and oxygen. The energy released by this fusion halts the contraction of the core and holds the star stable for about a million years. The center of the star now consists of a hot, dense shell of helium *(orange)* that encloses an even hotter and denser carbon and oxygen core *(green)*.

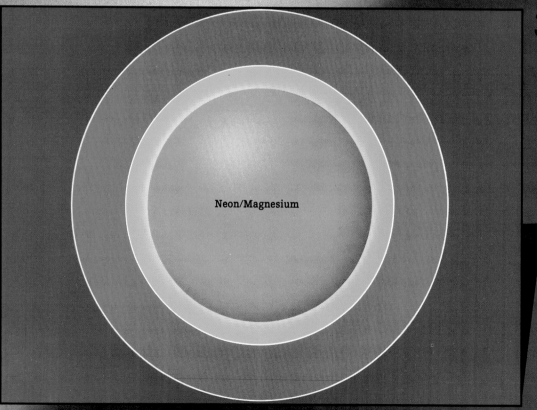

Neon/Magnesium

3 **One thousand years to go.** When most of the helium in the inner core has been used up, fusion energy once again cannot counter-act gravitational collapse, and the core begins to contract. From now on, the periods of contraction and fusion will alternate with ever-increasing speed as the star hastens toward the end. When gravitational contraction raises the temperature of the carbon core to 700 million degrees, fusion reactions begin converting carbon to neon and magnesium *(gold)*. Fusion at lower temperatures in layers surrounding the core continues converting helium to carbon and hydrogen to helium.

4 **Seven years to go.** When the temperature at the heart of the star's collapsing core reaches 1.5 billion degrees, neon atoms fuse to form more oxygen and magnesium *(yellow)*. The core begins to resemble an onion, its concentric layers of elements increasing in density toward the center.

Oxygen/Magnesium

5 **One year to go.** As the core temperature soars above two billion degrees, the most tightly compressed oxygen atoms fuse to form silicon and sulfur *(white)*. Surrounding these elements are concentric shells of oxygen, neon, carbon, helium, and hydrogen.

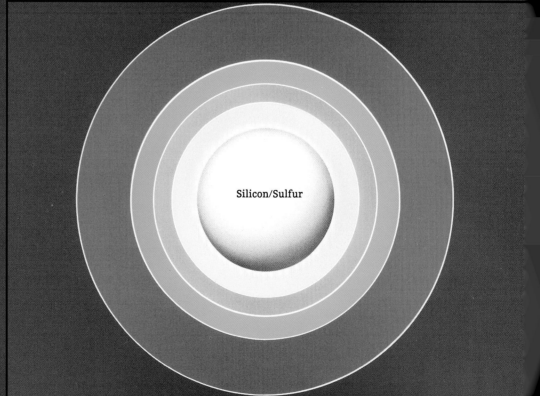

Silicon/Sulfur

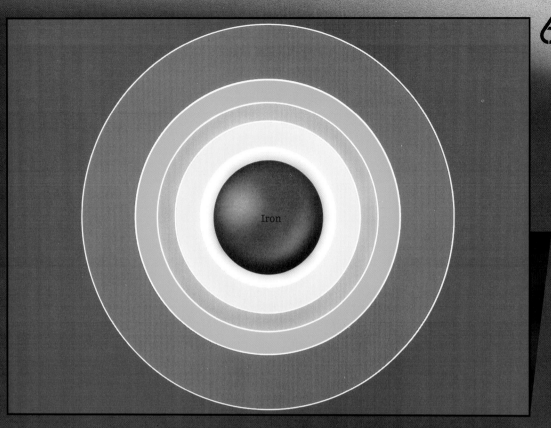

6 **A few days to go.** The mounting pressure of the collapse drives the temperature past three billion degrees, converting silicon and sulfur in the heart of the contracting core to a tightly compressed sphere of iron *(silver)* that measures approximately 1.44 solar masses. Because of iron's nuclear structure, which does not permit its atoms to fuse into heavier elements, this is the last reaction that can take place in the core.

Iron

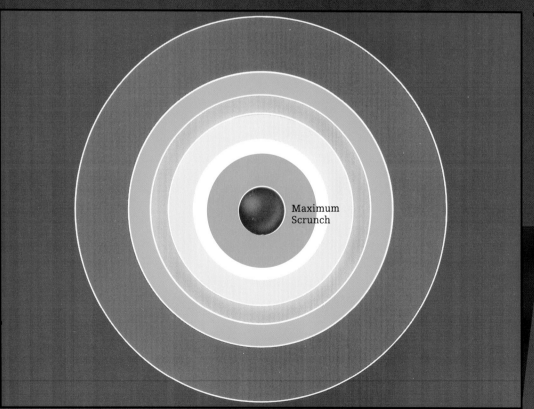

7 **Tenths of seconds to go.** When fusion reactions in the innermost core cease, the star begins the final phase of gravitational collapse. The iron heart of the star crushes in on itself at speeds approaching 45,000 miles per second, roughly one-quarter the speed of light. The core's temperature rises to 100 billion degrees as an Earth-size object is packed into a sphere just ten miles across. Iron nuclei are so compressed that they overlap and melt together. The incredible heat produces vast numbers of chargeless particles called neutrinos, which are temporarily trapped in the dense core. Matter in the core has now reached the point physicists call "maximum scrunch" and can endure no further compression. The repulsive force between the nuclei overcomes the force of gravity, and like a tensely coiled spring, the inner part of the iron core snaps back.

Maximum
Scrunch

8 **Milliseconds into the explosion.** The recoil of the central core hurls matter out from the heart of the star in an explosive shock wave that blasts through the silicon layer, heating it and fusing some of its nuclei into radioactive isotopes of nickel and other heavy elements. The explosion will continue to create new elements as it speeds out through the layers of the core.

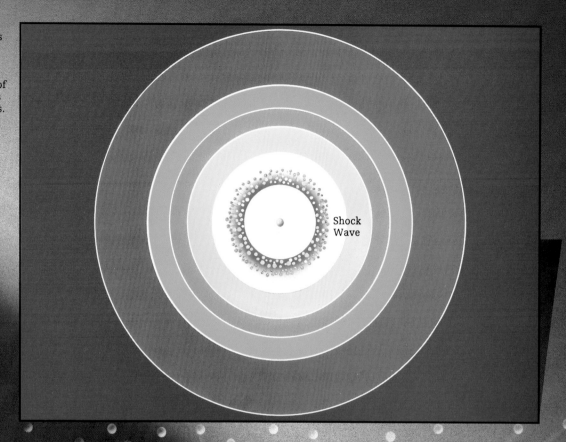

Shock Wave

9 **Seconds into the explosion.** The blast releases 99.5 percent of its energy in the form of neutrinos. Streaming easily through the star's outer layers, well ahead of the shock wave, the neutrinos are the first perceptible sign that the star is exploding. As the shock wave sends matter from the interior into space, all that remains is a small, superdense sphere composed almost entirely of neutrons—a neutron star.

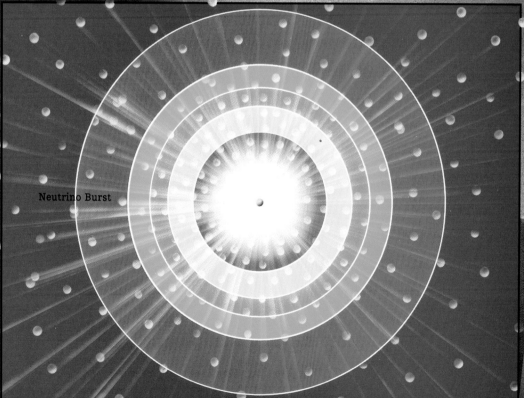

Neutrino Burst

Circumference before
Core Collapse

Neutron Star

10 Hours into the explosion. In the wake of the neutrino burst, shock waves erupt through the surface of the star, spewing several solar masses of newly created elements into space. (The perimeter of the star as it was before core collapse is outlined in red.) This expanding cloud of material will remain visible for thousands of years. The shock waves grow weaker as they move out from the center of the explosion, but they will compress interstellar matter for light-years in all directions, stimulating new cycles of stellar life and death.

1 Type Ia supernovae are thought to begin with two main sequence stars circling a common center of gravity. The more massive of the two *(upper left)* evolves faster, turning into a red giant while the other is still in its main sequence stage. As the giant's core fuses the last of its hydrogen to helium and begins to collapse, mounting internal heat forces the outer envelope to swell. The gas balloons out until it extends beyond the star's sphere of gravitational dominance and is captured by the gravity of the second star. Several solar masses of hydrogen then flow to the main sequence companion.

2 Hydrogen flooding from the first star may build up so quickly that the companion's gravity cannot hold it all. Some of the escaping gas then forms a cloud that cloaks both stars. This shared envelope drags on the stars and changes their orbits, bringing them closer together. The distance between the two stars decreases by as much as 90 percent, and their orbital movement creates an eggbeater effect that churns up the envelope, sending most of it flying out of the binary system.

Partners in a Fatal Dance

Type II supernovae such as SN1987A throw off enormous quantities of hydrogen. Other kinds of explosions are notably hydrogen-poor. To explain these enigmas, known as Type I supernovae, astronomers have developed theoretical models involving binary star systems—pairs of stars so close together that they exert substantial influence on one another's evolution. One model for a so-called Type Ia supernova, is illustrated here. In this scenario, the cause of the explosion is a runaway thermonuclear reaction triggered by one star's increase in mass as its gravity draws matter away from its companion.

The two stars that produce a Type Ia supernova each measure no more than eight solar masses at the beginning of their lives. Either partner, if on its own, would eventually fuse its core hydrogen into helium and progressively heavier elements, expand into a red giant, lose its outer hydrogen envelope, and

then contract to a dense white dwarf star. Composed primarily of carbon and oxygen measuring no more than 1.44 solar masses, the white dwarf's small mass would not be able to overpower the strong repulsive forces in its carbon interior. Core collapse thus would stop long before densities became great enough to ignite the carbon. Cooling gradually through a long old age, the white dwarf would dwindle to a spent stellar cinder.

The scenario is different, however, for stars orbiting a common center of gravity. The binary pair shown here and on the next two pages begin the main sequence portion of their lives separated by several astronomical units (the distance from Earth to the Sun). Once the companions reach the white dwarf stage of evolution, mutual attraction can cause them to merge into a single object whose mass is great enough to kindle a thermonuclear blast.

3 All that remains of the red giant after the loss of matter is a dense core that has fused to carbon and oxygen. Stripped bare, this core becomes a white dwarf, a star as massive as the Sun but no bigger than the Earth. The white dwarf and its main sequence companion, which now contains the only hydrogen in the system, continue to orbit their common center, but at much closer range. In time, the second star *(lower right)* reaches the red giant phase and produces a swollen hydrogen envelope.

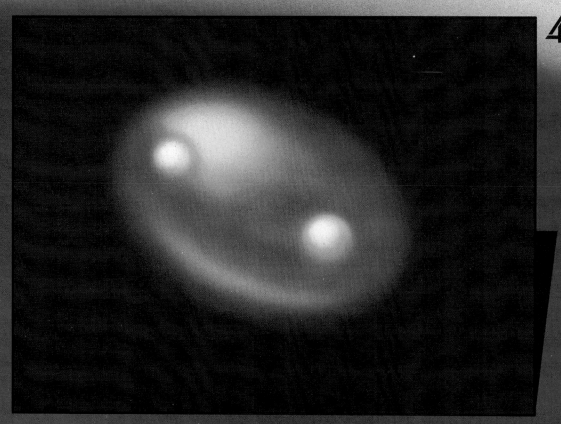

4 Eventually the second red giant star expands enough to lose gravitational control of its outer layers, and the cycle begins again. Hydrogen flows toward the white dwarf companion, leaving the evolving star with a core of helium and forming another shared envelope. This gas drags on the two stars, bringing them closer. Again their combined orbital action drives away most of the matter in the envelope, entirely depleting the system of hydrogen. Over time, the helium core of the evolving star converts to carbon and oxygen—a white dwarf—and the distance between the pair of white dwarfs continues to shrink.

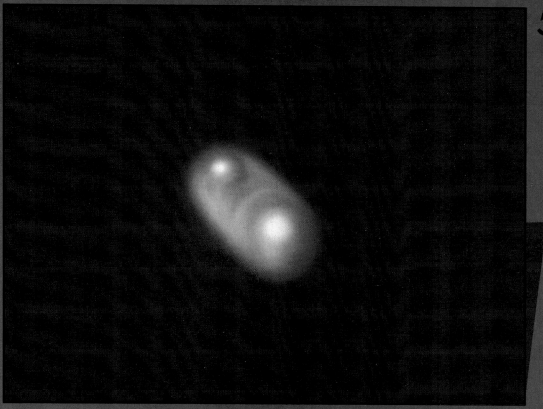

5 The two white dwarfs are now orbiting scarcely one million miles apart. Their gravitational interaction produces gravity waves that carry away some of their orbital energy, making their eventual merger inevitable. The smaller of the two stars, which is also the denser and more massive member of the pair, begins to strip matter from its companion, forming an envelope of carbon and oxygen.

6 Eventually the two white dwarfs circle so closely that, in effect, the stars merge, the mass of one accumulating on the surface of the denser companion. This added mass further compresses the receiving star, exceeding the critical limit of 1.44 solar masses and igniting a thermonuclear flame.

7 As the thermonuclear fire rips through the combined mass of the two stars, carbon and oxygen atoms are fused into a variety of heavy elements, some with strong radioactive emissions. As with the explosion of a Type II supernova, this stellar conflagration flings matter into space—but without even a trace of the hydrogen with which the stars began.

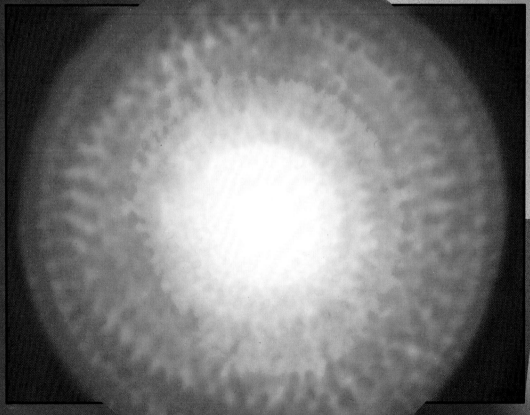

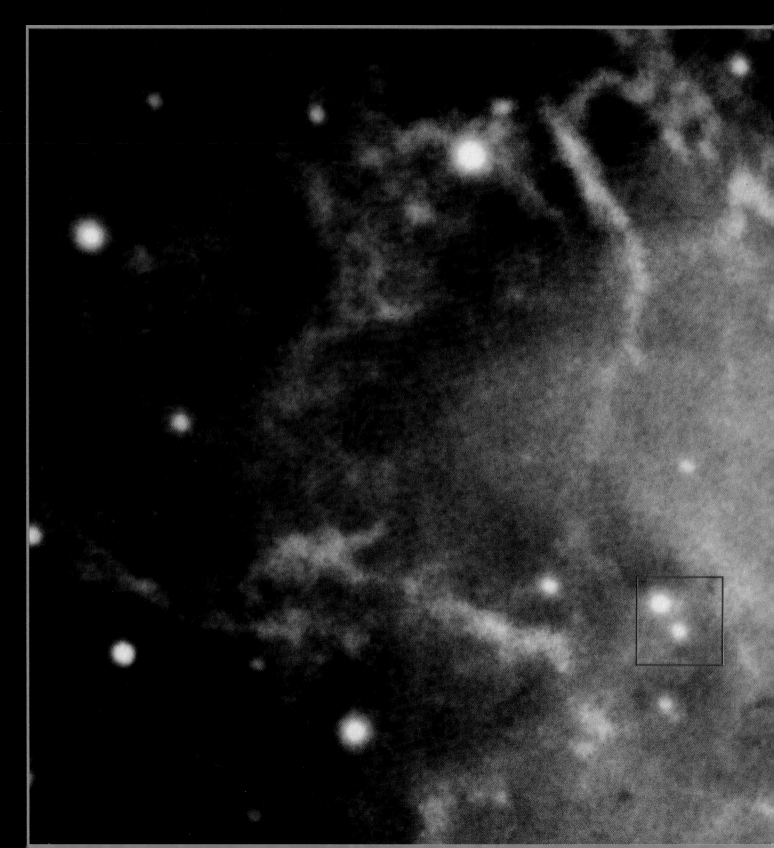

HOLES

A radiant swirl of gas and dust called the Crab nebula surrounds the remains of its progenitor—the supernova of 1054—which is now a rapidly spinning star known as a pulsar *(box, lower right star, and page 105).*

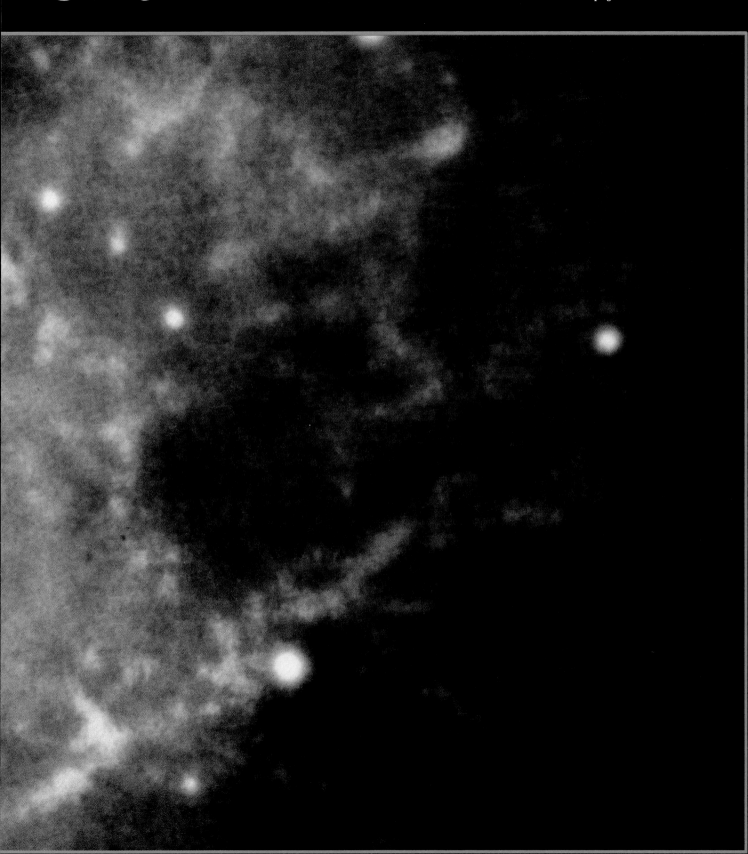

xploding stars, degenerate dwarfs, nuclear tunneling, and the uncertainty principle: By the 1930s the world of astronomy was reeling under a barrage of new theories and discoveries. It is little wonder that when a pair of U.S. scientists suggested yet another bizarre object, their peers revolted. The suggestion was almost timid, an "additional remark" at the end of a 1934 paper on supernovae. Its well-respected authors were Walter Baade of California's Mount Wilson Observatory and Fritz Zwicky of the California Institute of Technology. "With all reserve," they wrote, "we advance the view that a supernova represents the transition of an ordinary star into a neutron star, consisting mainly of neutrons. Such a star may possess a very small radius and an extremely high density."

They had good reason for such careful wording. The very idea of neutrons was new; the particle had been discovered only two years earlier by an English physicist, James Chadwick. With a mass slightly greater than that of a proton, its companion in the atomic nucleus, and having no electric charge, the neutron would play a major role in the subsequent study of nuclear physics. In 1934, however, scientists were just beginning to speculate on its function. Just as Baade and Zwicky must have feared, their paper in the *Proceedings of the National Academy of Sciences* evoked hoots of disbelief and outright ridicule. During the 1930s, even the existence of dense white dwarfs was in dispute. Few astronomers wanted to contemplate any object with the outrageous vital statistics of a neutron star. For instance, such a star could pack a mass greater than that of the Sun into a sphere only ten miles wide. The surface gravity of so dense an object would be 100 billion times stronger than that on the surface of Earth, enough to crush a human instantly into oblivion.

Ultimately, the paper was consigned to the dustbin by all but a handful of astronomers and physicists. Baade went quietly back to other studies. Zwicky, more contentious, continued to push the theory, but with little success. No one had ever observed a neutron star, and few expected to. Yet neutron stars were not the strangest objects to be suggested early in the twentieth century. By the time they were proposed, a few scientists had already realized that a supermassive dying star would not halt its collapse until it reached a point of infinite density. Like the neutron star, this incredible object—later known as a black hole—was ignored for years as one more wild-eyed speculation. Not until the 1960s did scientists employing radio and x-ray telescopes begin to observe certain strange signals from space that suggested these most science-fictional of stars were reality.

Physicist George Gamow was one of the few scientists who followed up on the neutron star idea. In 1937 he worked out the mathematical details of a process called neutronization. When the core of a supernova collapses, crushing in on itself at inconceivable speed, the enormous pressure forges neutrons by smashing protons and electrons together. Such neutron matter is almost incapable of further compression. Gamow proposed that a neutron star of the same mass as the Sun would be a sphere just six miles across. Two years later, two physicists at the University of California at Berkeley—the American J. Robert Oppenheimer (soon to be famous as director of the Manhattan Project, which built the first atomic bomb) and a Canadian graduate student named George Volkoff—defined the theoretical mass limits for neutron stars as lying between one-tenth and seven-tenths of the Sun's mass. Later, scientists would revise the higher limit upward; the precise number is still uncertain but is unlikely to exceed three solar masses.

Although important in hindsight, none of these early papers on neutron stars evoked much interest at the time. The problem was one of proof: It was all well and good to show that neutron stars ought to exist, but figuring out how to observe one was another matter. Such stars would emit some light from their surfaces, but not enough to detect at astronomical distances because they would be so pitifully small. For the next quarter of a century, neutron stars remained little more than an intriguing hypothesis.

So the matter stood until the mid-1960s and the heyday of radio astronomy. Radio telescopes, developed in the wake of World War II, added a new dimension to the sky by picking up signals from previously invisible objects and

In this sequence of optical images, the stellar remnant at the heart of the Crab nebula *(inset box, lower right star)* disappears and reappears as it pulses on and off at the rate of thirty times a second. From its magnetic poles, the pulsar radiates beams of energy that span much of the electromagnetic spectrum, including the radio and visible wavelengths.

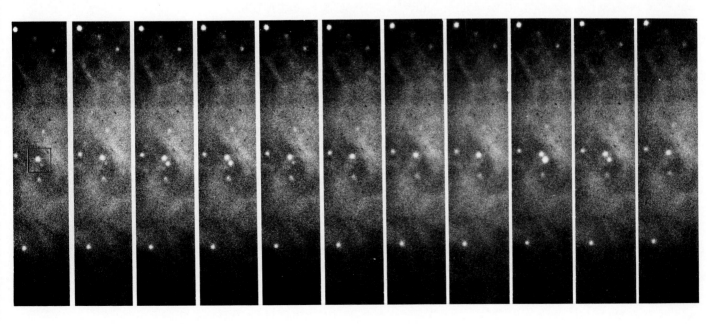

phenomena. Even the static-causing solar wind, a hurricane of charged electrons spewing from the Sun's surface, proved helpful because it made cosmic radio signals "scintillate" in much the same way that starlight twinkles as it passes through Earth's atmosphere. Astronomers found out that the degree of scintillation depended in part on the diameter of the radio source: The bigger the source, the less its signal varied. Since information about the true size of celestial objects was often hard to come by, Cambridge students under the direction of British astronomer Antony Hewish built a radio telescope especially to study scintillation. By 1967 Hewish, forty-three, was already launched on an illustrious career at Cambridge, where he taught physics; he had no time for the drudgery of scrutinizing radio signals. When the new telescope went into operation in July of that year, the job fell to one of his graduate students, a spunky twenty-four-year-old named Jocelyn Bell. This daughter of an architect may have demonstrated the necessary tenacity for the task by spending the previous two years wielding a sledgehammer in the telescope construction project.

The job was certainly not one for the easily discouraged. Whenever the telescope picked up a signal, a spike would appear on chart paper that scrolled continuously past an automatic pen. Different radio sources (whether a solar flare, a star in the Milky Way, or a distant galaxy) produced different spike patterns. Unfortunately, the telescope did not discriminate between celestial and man-made signals—the simple act of turning on a car ignition registered as easily as radio waves from the remote stellar objects known as quasars. Over time, Bell became adept at distinguishing among various types of signals recorded on the chart. Then, in October, an odd pattern appeared.

"Six or eight weeks after starting the survey," Bell later reported, "I became aware that on occasions there was a bit of 'scruff' on the records, which did not look exactly like a scintillating source, and yet did not look exactly like man-made interference either." The peculiar grouping of spikes took up only half an inch of the ninety-six feet of chart paper that was the telescope's daily output, and it was absolutely unique. Checking the records, Bell found that the pattern had first appeared in August but had gone unnoticed. To her frustration now, the scruff vanished for several weeks before abruptly reappearing in the middle of the night on November 28.

"As the chart flowed under the pen," Bell said, "I could see that the signal was a series of pulses, and my suspicion that they were equally spaced was confirmed as soon as I got the chart off the recorder." The signals were one and one-third seconds apart. Bell telephoned Hewish immediately. Later, when the two measured the signals more precisely, they found the spikes repeated at extremely regular intervals: Without fail, the time between signals measured exactly 1.3373011 seconds. Furthermore, the scruff appeared unique not only in the precision but also in the rapidity of its pulses. The only other celestial object with regular fluctuations was a so-called variable star—a star that waxes and wanes in brightness. And the fastest known period for a variable star was thousands of times slower than Bell's object.

British astronomer Jocelyn Bell, who first discerned radio signal patterns that proved to be the pulsing of rapidly spinning neutron stars, or pulsars, stands in a field of antennas she helped erect near Cambridge, England, in the late 1960s. More than a thousand of these wooden poles, strung with one and two-thirds tons of copper wire, made up a radio telescope that covered about four and a half acres. Bell found the extremely regular patterns while examining several miles of graph paper on which the telescope's observations were recorded.

Because the signals came from beyond the Solar System, Hewish speculated that the sender was some extraterrestrial intelligence. For a while, Bell and Hewish referred to the signals as "LGMs"—for Little Green Men. Bell later claimed that neither she nor Hewish took this hypothesis very seriously, although, she said, it had to be considered: "Radio astronomers are aware that they would probably be the first people to come into contact with other civilizations." So Hewish thought of a test. He assumed that any extraterrestrials would be living on a planet orbiting a star. As the planet moved toward or away from an earthbound observer, radio signals coming from it would be bunched up or spread out in the familiar phenomenon known as the Doppler shift.

No Doppler shift turned up when Bell examined the record, weakening the case for LGMs but not ruling them out entirely. The day before leaving on Christmas holiday, Bell visited Hewish to discuss ways of announcing their discovery without setting off a media circus. They could think of none. She recalled returning home feeling "very cross" about trying to complete her Ph.D. dissertation while "some silly lot of little green men had to choose my aerial and my frequency to communicate with us." After dinner Bell returned to the laboratory to analyze the previous twenty-four hours of chart output. Her pulse leaped when she found more scruff, this time in the constellation Cassiopeia—an entirely different part of the sky. She checked the record for other days and recognized the same pattern. The patch of sky in which this new bit of scruff appeared would be passing overhead in a few more hours— she might be able to catch it. But the night was bitterly cold, and under such conditions the telescope's receiver often got cranky and failed to record any but the strongest signal. In desperation, Bell breathed on the device to warm it up, flicked switches, cursed it—and got it to work properly for five minutes. The young astronomer knew it was the right five minutes as she watched the pen scratching the unmistakable signature of the Cassiopeia signal.

Two signals arriving from widely separate parts of the sky put the LGM hypothesis to rest, at least in Bell's mind. The chances of two civilizations trying to contact Earth at the same time were statistically so improbable that it could be safely discarded. The now much happier graduate student left the recording on Hewish's desk and dashed off for her Christmas holiday.

After finding two more pulsating sources in January of 1968, Bell stopped observing to begin writing her dissertation on the effect of solar wind on radio signals. The four signals were mentioned only in an appendix. But the specter of little green men refused to go away. After Hewish announced the discovery of pulsating radio sources—soon abbreviated to pulsars—in February, the British tabloid press jumped on the LGM hypothesis. That the astronomer who found them was a young and attractive female made the story irresistible. Reporters wanted to know if Bell was taller or shorter than Princess Margaret. They badgered her about her love life. She endured endless photo sessions in various contrived poses; one photographer asked her to wave her arms hysterically as though she had just made contact with extraterrestrials.

The public lost interest in pulsars once it realized they were not the product of some alien intelligence. But for the astrophysical community they remained a hot new topic. Over the next couple of years, articles poured in to scientific journals as new pulsars were found and as theorists attempted to explain them. What exactly were these baffling objects, and how did they generate their rapid-fire bursts of energy? Some, Hewish and Bell among them, thought they might be either white dwarfs or those long-ignored entities, neutron stars. Somehow, these scientists proposed, the dense little stars literally pulsed, producing radio waves as they expanded and contracted.

Meanwhile, official laurels for discovering pulsars went to Hewish. In 1974 he was awarded a Nobel prize in physics. The fact that Bell did not share in science's most prestigious honor rankled many in the field, including Britain's Fred Hoyle, who had contributed so much to cosmology and astrophysics: In a 1975 letter to the *Times* of London he blasted her exclusion as a scandal. Bell declared Hoyle's criticism "a bit preposterous"; Hewish for his part replied that Bell had been working under his direction and would have been negligent if she had missed the odd bits of scruff leading to the discovery, even in the absence of specific instructions to look for them.

Not everyone was satisfied with this explanation, but little could be done. By 1974 Bell had gained her Ph.D. and had gone on to work as a researcher at England's University of Southampton. Although she was not granted Nobel recognition for her discoveries, she did win other prestigious science prizes over the years, and her place in science is assured. Her achievement, noted Hoyle, "came from a willingness to contemplate as a serious possibility a phenomenon that all past experience suggested was impossible."

A STELLAR DYNAMO

Hewish's notion of pulsars as throbbing stars was soon superseded in 1968 by a more likely theory from Thomas Gold of Cornell University, an astronomer known for his unorthodox ideas. He proposed that a pulsar is a neutron star that emits a narrow beam of radio energy as it spins rapidly on its axis. The star's dizzying rotation rate follows the law of conservation of angular momentum: Just as a figure skater spins faster when she pulls in her arms, a star collapsing from its red giant phase spins faster because the rotational velocity of its larger self is now operating in a drastically shrunken body.

The mechanism for generating the pulsar beacon was more difficult to explain (and is still not fully understood). In Gold's scenario, the pulsar's rapid spinning acts as a huge dynamo, accelerating charged particles on its surface to nearly the speed of light and focusing them into a narrow beam along the magnetic poles. As the particles spiral along the beam, their energy is converted to various forms of electromagnetic radiation. If the star's magnetic poles do not coincide with its axis of rotation, the beams sweep around in a circle, much like the beams of a lighthouse. When the Earth happens to lie along the path of those beams, observers will see the pulsar as a flash of energy each time the beam touches the planet *(pages 111-115).*

Gold's version of pulsars became the generally accepted model. However, it was soon apparent that there might be many variations on the norm. In 1968 radio astronomers in Green Bank, West Virginia, found a rapid pulsar at the center of the Crab nebula, the ghostly remnant of the supernova recorded by Chinese chroniclers in 1054. Scientists were jubilant; here was early support for their theories, since the pulsar was located just where one would expect a neutron star to be in the aftermath of a supernova. As predicted, it was slowing down ever so slightly as its rotational energy was converted into particles and radiation that flooded out into the nebula. But even more exciting was the discovery, made at Arizona's Kitt Peak Observatory in 1969, that the pulsar was flashing in the range of visible light. By synchronizing their optical recording system with the thirty-times-a-second pulse of the object, astronomers were able for the first time to capture the flickering light beaming from a distant neutron star.

Since 1969 more than 400 pulsars have been discovered. Perhaps the most interesting is one in the constellation Aquila (the Eagle), whose pulse rate rises and falls in a regular cycle of seven hours and forty-five minutes. It was discovered in 1974 at the giant Arecibo radio telescope in Puerto Rico by University of Massachusetts graduate student Russell Hulse, who hypothesized that the pulsar was orbiting an unseen companion (perhaps another pulsar, but one whose radio beam was pointed away from Earth so it could not be observed). If so, Hulse concluded, the pulsar's period only appeared to vary. On each orbit, the neutron star's movement relative to Earth was first away from, and then toward, the planet. As the movement changed, the wavelength of the radio signals received on Earth was lengthened or shortened by the Doppler shift. Hulse and his supervisor, Joseph Taylor, realized that certain properties of the Aquila binary system made it a possible testing ground for relativity. Einstein's general relativity theory, published in 1915, says in effect that gravity is not a force as Newton conceived it but the bending of time and space in the presence of a massive object. The equations of general relativity make predictions about the gravitational influence of one body on another that differ from predictions using older Newtonian physics. The differences are small, however, and scientists cannot detect the variances unless the objects in question are at least as massive as a star.

Ordinary binary star systems might serve as test beds except that the mutual gravitational attraction between normal stars causes deforming bulges that affect orbital motion. By contrast, systems involving neutron stars combine enormous mass with extreme compactness, thus eliminating the confusing effects inherent in normal binary systems. Scientists keen to test relativity were therefore elated to find the Aquila binary pulsar, not only because of its short period and the compactness of each of its partners, but because the incredible regularity of its signals made it the best timepiece in the universe for measuring gravitational effects between two stars.

Among several testable predictions was one regarding the existence of gravity waves, ripples in the geometry of space around massive objects. Einstein said that gravity waves radiating from a massive object should carry away a tiny fraction of its energy. In a binary pulsar, this loss in energy would result in an ever-tighter orbit and a corresponding increase in orbital speed. Thus, it would take less time for the pulsar to complete an orbit—a change that can be calculated using Einstein's equations. Taylor and his students monitored the Aquila pulsar for five years. In 1978 they announced that the star's orbital period was changing at exactly the rate predicted. The speed-up attributable to gravity waves was absurdly small—on the order of one second every 10,000 years—and its discovery was only possible because the pulsar's extraordinary precision made the tiniest changes in its period obvious.

SINGULARITIES
When Einstein developed relativity theory, it took him ten years to work out the concepts and to describe them in a forbidding form of mathematics called tensor calculus. The great physicist himself produced only approximate solutions to his own equations, and the math involved still daunts even the best scientific minds. However, the difficulty of general relativity did nothing to deter an energetic contemporary astronomer and theoretical physicist named Karl Schwarzschild, director of the Astrophysical Observatory at Potsdam, now in East Germany. Schwarzschild, whose son Martin would make important contributions to the study of late stellar evolution *(pages 74-78)*, had a practical bent; among other things, he pioneered new methods of studying spectra. But his aptitude for the theoretical was evident early on. By the age of sixteen, he had already published a paper on the movement of heavenly bodies. When Einstein's articles on general relativity appeared in 1915, Schwarzschild was one of the first to see their importance.

Unfortunately, he was not in the best position to work on astrophysics. Germany was in the throes of World War I, and the astronomer, a fervent patriot, had taken leave from his post to enlist in the army, although he was more than forty years old and could easily have avoided service. By the time he read Einstein's writings, he had already seen action in Belgium and France and had just been transferred to the Russian front. Enthralled by what he called the "purity" of general relativity, he began at once to find an exact

RADIO PULSES FROM COSMIC BEACONS

First observed by radio telescopes in the 1960s, pulsars are energy sources that emit radiation in bursts of extraordinary regularity. Their signals repeat at intervals ranging from several seconds to mere thousandths of a second. So precise were the first pulsars' repeat times that their discoverers were half tempted to attribute them to intelligent extraterrestrial beings. Ultimately the mysterious signals were found to be something equally sought after: proof of the existence of neutron stars, first predicted in the 1930s as the cores that would be left after massive stars exploded as supernovae *(pages 89-101)*.

The dense compression of matter inside a neutron star gives it a magnetic field that is a trillion times as powerful as the field of an ordinary star. This magnetism, in combination with the star's extremely rapid rate of rotation, produces a kind of dynamo effect. A spinning neutron star hurls electrically charged electrons and protons from its surface. Spiraling along the star's magnetic lines of force at speeds approaching that of light, these particles emit electromagnetic energy of various types, including radio waves, x-rays, and gamma rays.

According to one theoretical model, the energy fans out from each of the neutron star's magnetic poles like two powerful beams, which the star's rotation transforms into a repeating beam like that of a lighthouse. The model accounts for most of the more than 400 pulsars so far detected, but—as illustrated on the following four pages—the galaxy is host to examples that wriggle out of this ingenious mold and demand some special explanations.

HEAVYWEIGHT REMNANT OF A SUPERNOVA

A neutron star is the smoking gun at the scene of a supernova explosion. The immense pressure and heat in the supernova's iron core at the moment of the explosion create a neutron star by forcing oppositely charged electrons and protons so close together that they fuse to become neutrons. Inside an iron shell only a few hundred feet thick, these neutrons pack together into matter so dense that a single teaspoonful could weigh a billion tons. Though no more than perhaps ten miles in diameter, the neutron star has a gravitational field at its surface 100 billion times as strong as that on the surface of the Earth.

The supernova explosion that forms a neutron star also expels vast numbers of nearly massless particles called neutrinos, as well as an expanding cloud of dust and gas that can remain visible for thousands of years.

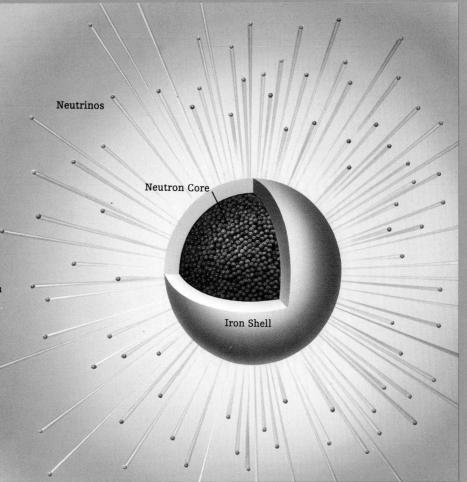

Neutrinos

Neutron Core

Iron Shell

A Celestial Lighthouse

The simplest model of pulsar behavior, depicted on this page, envisions the object as a dense, rotating sphere of neutrons that emits conical beams of radio waves from its two magnetic poles. If the magnetic axis of the star coincided with its axis of rotation, no pulsating effect would result; the emitted energy would simply radiate steadily. However, because the star's rotational and magnetic axes do not coincide, each beam rotates as the pulsar spins, like the rotating beam of a lighthouse. Observers on Earth can only detect the emissions when the axis of the cone points directly earthward: Radio telescopes will record a pulse each time a beam of radiation sweeps across them. Astronomers can calculate how fast the pulsar is rotating from the period between pulses.

Electrons and protons in the grip of the neutron star's powerful magnetic field are accelerated to near the speed of light along spiral pathways, producing cones of radiation that beam out in line with the magnetic axis of the star.

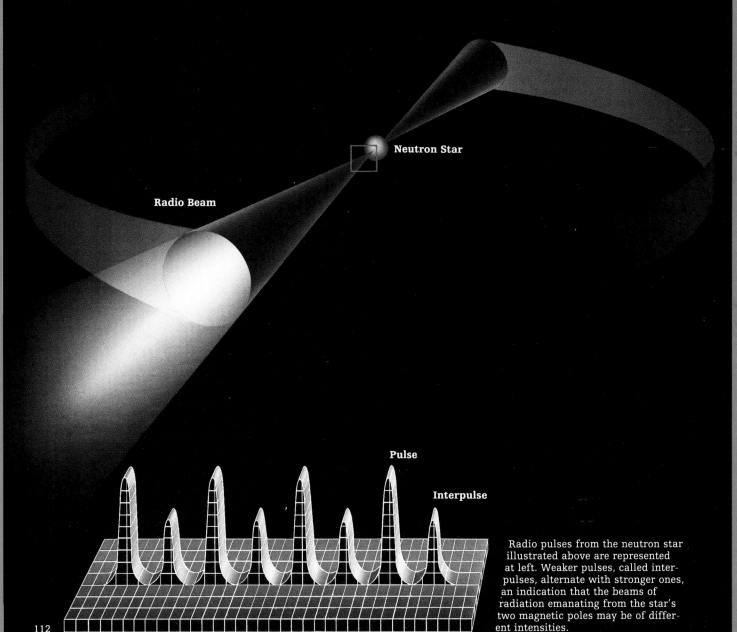

Neutron Star

Radio Beam

Pulse

Interpulse

Radio pulses from the neutron star illustrated above are represented at left. Weaker pulses, called interpulses, alternate with stronger ones, an indication that the beams of radiation emanating from the star's two magnetic poles may be of different intensities.

QUINTUPLE SIGNALS FROM A LONE STAR

One of the most unusual pulsars ever to have been detected produces a five-fingered radio signature. Astronomers theorize that five separate conical beams are emanating from the magnetic poles of PSR 1237 + 25, as the anomalous object is known. The result is a complex group signal that is created as the individual beams rotate together about the star's magnetic axis. The rotation affects the pulsar's signal by causing the component pulses to fluctuate in strength periodically.

Scientists assume the pulsar's radiation is generated according to the widely accepted theoretical model—by protons and electrons spiraling around the star's magnetic field. But they have not yet been able to explain the specific properties that might produce its multibeam structure.

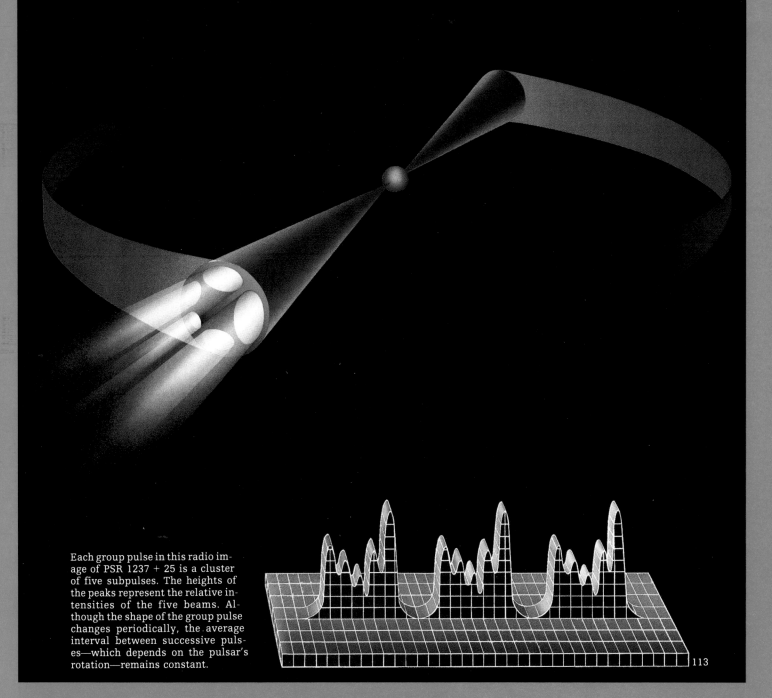

Each group pulse in this radio image of PSR 1237 + 25 is a cluster of five subpulses. The heights of the peaks represent the relative intensities of the five beams. Although the shape of the group pulse changes periodically, the average interval between successive pulses—which depends on the pulsar's rotation—remains constant.

PULSARS IN PAIRS

The radio signature from pulsar 1913 + 16 suggests that it is part of a binary system that includes not one but two neutron stars. Its companion is not detectable from Earth, presumably because its radio beams never sweep across this part of the galaxy. However, changes in the signal of PSR 1913 + 16 form a pattern that indicates it is orbiting another object. In the phenomenon known as the Doppler effect, the pulsar's bursts of radiation increase in frequency when the star's orbit carries it toward Earth and decrease as the orbital path carries the star away.

Perhaps more exciting to astrophysicists is that the pulsar's signals seem to reveal that both neutron partners are losing orbital energy and are moving closer together. Scientists speculate that this loss of energy is caused by the generation of gravity waves—a not-yet-detected form of radiation that Einstein's general theory of relativity predicts should be emitted by two orbiting masses.

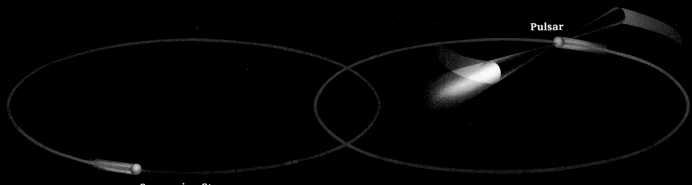

Pulsar

Companion Star

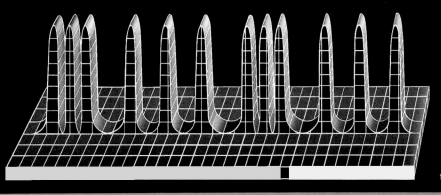

The radio reading of PSR 1913 + 16 shows cyclical variations in the frequency of its pulses, a sign that the star is orbiting the gravitational center of a binary system. As the pulsar's orbital path brings it toward Earth, the time between pulses decreases; as the star moves away, the time between bursts increases.

114

THE CREATION OF A MILLISECOND CLOCK

Pulsar 1937 + 21 can lay claim to two astronomical distinctions. Not only does it emit bursts of radiation 600 times a second, making it among the half-dozen fastest pulsars so far discovered, it is also one of the most precise timekeepers in the universe. The interval between each of its pulses is 1.6 milliseconds, recurring with a regularity that may surpass the accuracy of the best atomic clocks.

The pulsar is something of an enigma. Because astronomers have not detected a cloud of supernova debris around PSR 1937 + 21, they believe that the neutron star must be old—the core of an explosion that dispersed long ago. But this age makes its very fast rate of spin a puzzle, since most observations suggest that neutron stars slow down as they age. Thus scientists speculate that the pulsar was once part of a binary system that included a conventional star. As the immense gravity of the neutron star dragged matter away from its companion *(right, top)* the neutron star began to spin faster. The companion then expanded into a red giant *(middle)* and eventually exploded as a supernova *(bottom)*, leaving PSR 1937 + 21 with a neutron star companion whose radio emissions are not detectable from Earth.

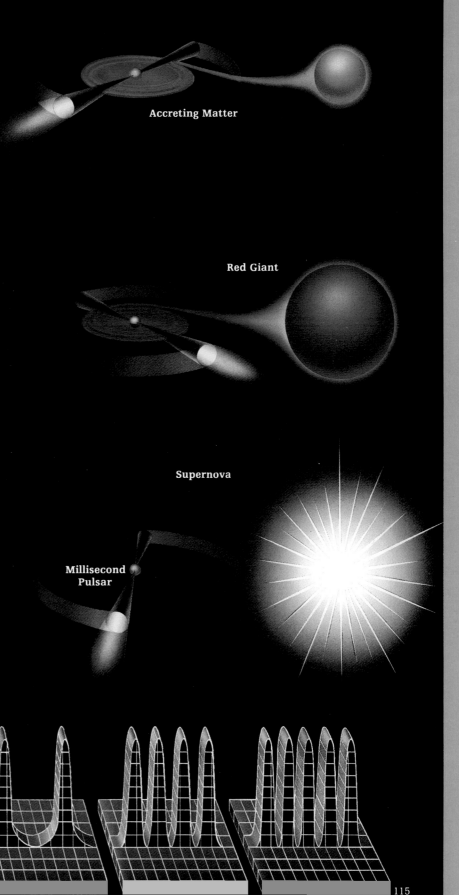

Accreting Matter

Red Giant

Supernova

Millisecond Pulsar

The graph at right represents the presumed history of the increasing pulse rate of PSR 1937 + 21. The sections of the graph reflect differences in the pulsar's signal at the three stages depicted above. As the neutron star spins faster, the frequency of pulses increases and the period between them decreases.

solution to its equations. But not only was he hampered by the hazards and disruptions of war, he also came down with pemphigus, a painful and debilitating disease of the skin and mucous membranes. Nevertheless, when not employed in calculating trajectories for long-range missiles at field headquarters, he worked on the mathematics of relativity. Two months after contracting the disease, he was sent home. He completed his work shortly before his death from pemphigus in May 1916.

The paper Schwarzschild wrote, innocuously titled "On the Field of Gravity of a Point Mass in the Theory of Einstein," was published later that year. It would become one of the foundations of modern relativity studies. Schwarzschild's solution to the Einstein equations had led him inexorably to the concept that if an object—no matter what its original size—were sufficiently compressed, it would possess a gravitational force so strong that not even light could escape it. He had realized that general relativity theory allows a mass to be squeezed into an infinitely small space. Scientists now refer to such a zero-volume mass as a singularity; Schwarzschild called it a point mass. His calculations also showed that around any point mass lies a spherical gravitational boundary called an event horizon. Inside this boundary, nothing can ever escape. The physicist worked out an equation that gave the size of any object's event horizon, a figure now known as a Schwarzschild radius. The Sun's Schwarzschild radius is 1.75 miles. For the Earth it is a third of an inch: If the Earth were compressed into a singularity, the sphere of its event horizon would be about the size of a small marble.

Despite its radical conclusions, Schwarzschild's paper was regarded as no more than a theoretical curiosity. Scientists would not grasp its implications for stellar evolution until after fifty years of progress in particle physics and astronomy. In any case, the very idea of a singularity troubled many, including Einstein, for the notion of infinite density flew in the face of experience. Nature, after all, is finite; things, whether elephants or protons, have boundaries and can be weighed and measured.

Yet the unsettling notion was given further credence in 1939, when Robert Oppenheimer followed up the paper on neutron stars he had published earlier in the year with a second article showing that a star with sufficient mass could indeed be squeezed by its own gravity into a dimensionless point. Oppenheimer wrote his paper with Hartland S. Snyder, another brilliant graduate student. The equations laid out by the two scientists demonstrated with irrefutable logic that a star of sufficiently great mass at the end of its existence would continue collapsing beyond the white dwarf and neutron star stages into a singularity. If the upper limit for neutron stars was seven-tenths of a solar mass, as Oppenheimer had earlier figured, then the minimum measure for singularities was somewhere in that neighborhood, or about one solar mass. (Subsequent theorists have decided that in order for a star to collapse to a singularity it must leave the red giant phase with at least two or three solar masses.)

Thus, by the eve of World War II, theorists had set the thresholds for the

With his work on so-called mini black holes, Cambridge University theoretical physicist Stephen Hawking has tackled the daunting task of unifying the two major aspects of physics—relativity theory and quantum mechanics—despite being confined to a wheelchair and unable to either write or speak.

three forms of stellar remnants: white dwarfs, neutron stars, and singularities. The work of Schwarzschild, Oppenheimer, and Snyder had proved—mathematically, anyway—that singularities could exist. However, by definition there was no way to observe them, so astronomers took little notice. Over the next twenty years, infinitely dense stars remained the exclusive province of a few theoretical physicists.

IN SEARCH OF COSMIC ENGINES

By the 1960s observed reality had begun to catch up with theory in unexpected ways. Astronomers came to realize that the universe seethed with unbelievably violent forces. Conventional physics was woefully inadequate to explain radio waves blasting from the center of the Milky Way or quasars burning with awesome fury at the edge of the universe. The most likely candidate for producing such vast outpourings of energy seemed to be matter spiraling around a supermassive singularity.

In 1963 the notion of singularities as cosmic engines was proposed by two of the most senior and respected men in the physical sciences, Oppenheimer and Britain's Fred Hoyle. Astronomers were at last ready to listen. Before the end of the decade, Princeton University physicist John A. Wheeler would coin "black hole" to describe the objects previously known by less evocative names such as "collapsar" and "frozen star." As awareness of black holes spread from scientific circles into the larger community, the idea stirred the popular imagination, and magazines ran articles about "The Darkest Riddle of the Universe" and even "The Blob That Ate Physics." But public fascination with black holes was minor compared with the excitement they were generating among theoretical physicists. As the 1960s ended, a new generation of thinkers began to ponder the nature of black holes and to reach even more startling conclusions about them. The undisputed leader of black hole theorists was an elfin, shaggy-haired Englishman named Stephen Hawking.

While still in his twenties, Hawking had made a name for himself in the rarefied realm of relativity theory, and his ideas—about such esoteric phenomena as exploding black holes and "wormholes" burrowing through space-time—would shake the foundations of scientists' concepts about mass, energy, and gravity. Moreover, Hawking the man was no less remarkable than his work. At the age of twenty, while a graduate student at Cambridge, he was diagnosed as suffering from the early stages of amyotrophic lateral sclerosis, a crippling and incurable nerve disorder also known as Lou Gehrig's disease, after the American baseball player who died from it. His illness, which destroys the nerves controlling voluntary movement, has put him in a wheelchair for the rest of his life and gradually reduced his voice to a stream of guttural sounds, so that he has come to converse by tapping on the keys of an electronic voice synthesizer. His mind, though, remains unimpaired.

Hawking's first major contribution came in 1971, when he showed that billions of tiny black holes may have been created in the very early universe. These mini black holes, each no bigger than an atomic nucleus but containing the mass of a mountain, could only have been formed in the enormous densities existing a fraction of a second after the Big Bang, the primordial explosion in which the universe is believed to have begun some 10 to 20 billion years ago. Three years later he struck out even farther into new territory when he published a paper announcing that black holes could erode over time, evaporating and even exploding as their mass leaked into space. This notion turned Einsteinian physics upside down. Conventional wisdom had it that anything caught within the event horizon of a black hole would have to exceed the speed of light to escape, and according to the rules of relativity this was physically impossible.

Hawking suggested another way for an object to free itself from a black hole's gravity prison. By applying certain laws of quantum mechanics to the physics of black holes, he decided that particles could gradually leak out over the event horizon. Quantum theory, which describes the behavior of matter on the subatomic level, predicts that pairs of elementary particles, matter and

antimatter, can appear in unpredictable places and then annihilate each other. According to Hawking's calculations, the extreme gravitational force of a black hole could produce such pairs just outside its event horizon. If so, one particle might be dragged in and the other escape. The hole would gradually lose the gravitational energy attached to such particles, and eventually it would explode. The process would take about 10 billion years, Hawking figured. Since the universe is at least 10 billion years old, and since billions of mini black holes may have formed at the beginning of the universe, it was possible—indeed probable—that, even now, cosmic explosions are occurring throughout the vastness of space. Hawking found this conclusion so perturbing that at first even he refused to believe it. Before publishing his paper he spent months poring over his equations, seeking some critical mistake. But the mathematics appeared to be flawless.

Moreover, theory suggested that these blasts could be detected. Mini black holes might generate high-energy gamma rays that could be perceived by specially equipped satellite telescopes. Gamma ray bursts had been seen by the late 1980s, and although they did not perfectly fit the description of radiation from mini holes, many relativists remain both convinced that mini black holes exist and hopeful of eventually catching sight of their passing.

Some scientists took black hole speculation further still. If black holes in some sense punch through the fabric of space-time, an object falling into one might emerge in some other part of the universe—or in another universe altogether. These hypothetical tunnels came to be called "wormholes," and not surprisingly they were seized upon by science-fiction writers and futurists as a potential shortcut through the vast distances of space (sidestepping the problem of an astronaut surviving a plunge into the crushing gravity of a black hole). Many physicists were skeptical, pointing out that the idea of both wormholes and exploding black holes derived from untestable assumptions. Undaunted, others thought of ways to employ the insatiable objects: A mini black hole might be captured, placed in orbit around Earth, and used to generate electricity; in other parts of the universe, advanced civilizations might already be clustered around black holes, just as early civilizations on Earth were built around rivers. A black hole orbiting the planet would also make the ultimate trash compactor for humanity's accumulating refuse.

For most astronomers, though, the joy of indulging the imagination paled beside the thrill of discovering the real thing. During the 1970s, x-ray astronomers began to feel that excitement when they turned their attention to a mysterious object in the constellation Cygnus.

TO SEE THE INVISIBLE

In December 1970, scientists launched a satellite-borne x-ray telescope from the coast of Kenya. Named *Uhuru*, the Swahili word for "freedom," the satellite was the brainchild of Riccardo Giacconi, an Italian-born astronomer and pioneer in x-ray technology. Because short-wavelength x-radiation does not pierce Earth's atmosphere, the telescope would give astronomers their

In the northern constellation Cygnus, an intense x-ray source designated Cygnus X-1 *(inset, blue area)* is believed by many astronomers to be a black hole. Invisible to optical instruments, Cygnus X-1 is detectable because it orbits a visible companion star, a blue supergiant named HDE 226868 *(left, red arrow)*. Gases drawn from the companion, heated to millions of degrees as they spiral into the black hole, emit energy at x-ray wavelengths.

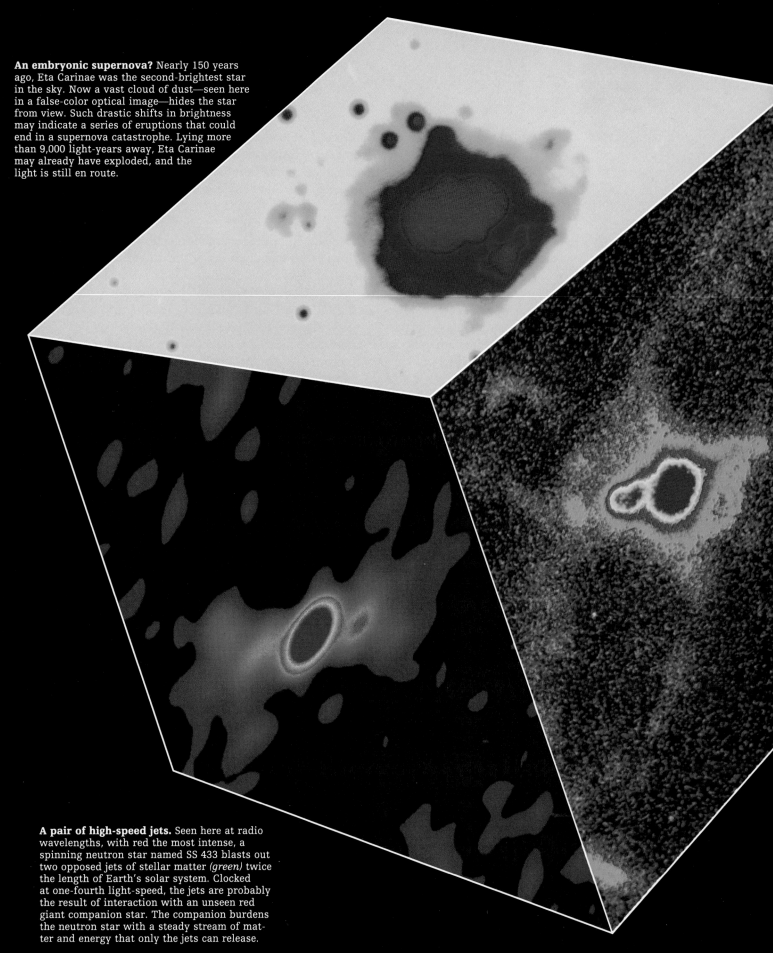

An embryonic supernova? Nearly 150 years ago, Eta Carinae was the second-brightest star in the sky. Now a vast cloud of dust—seen here in a false-color optical image—hides the star from view. Such drastic shifts in brightness may indicate a series of eruptions that could end in a supernova catastrophe. Lying more than 9,000 light-years away, Eta Carinae may already have exploded, and the light is still en route.

A pair of high-speed jets. Seen here at radio wavelengths, with red the most intense, a spinning neutron star named SS 433 blasts out two opposed jets of stellar matter *(green)* twice the length of Earth's solar system. Clocked at one-fourth light-speed, the jets are probably the result of interaction with an unseen red giant companion star. The companion burdens the neutron star with a steady stream of matter and energy that only the jets can release.

Close companions. Every forty-four years, the stars of the binary system R Aquarii approach each other so closely that one—a stellar giant seen as the larger red spot in this false-color view—pours gases toward its white dwarf partner, not visible here. The sudden transfer triggers gaseous outbursts, some of which may have created a nest of outlying clouds and the red-coded jet at the giant star's left.

A future solar system? With much of its own light excluded by a high-tech cross hair, Sun-like Beta Pictoris reveals an extensive, dim disk of material in this electronic image. Astronomers picture the disk—ten times the diameter of the Solar System—as a plane of tumbling ice and rock fragments that may one day coalesce into planets. Since something has apparently swept out the disk's inner region, a world—or worlds—may already have formed.

An Array of Oddities

Every star is an individual, but some are truly odd, departing radically from typical stellar behavior. Among the sampling of eccentrics shown here, one *(far left)* squirts out streams of hot plasma at more than 170 million miles per hour, a feat unmatched by any other known star. Another *(below)* nearly winks out from time to time for periods lasting from several months to several years, then gradually reappears as before. Hundreds of anomalies like these have been spotted; for the most part, the reasons behind their nonconformist ways can only be guessed at.

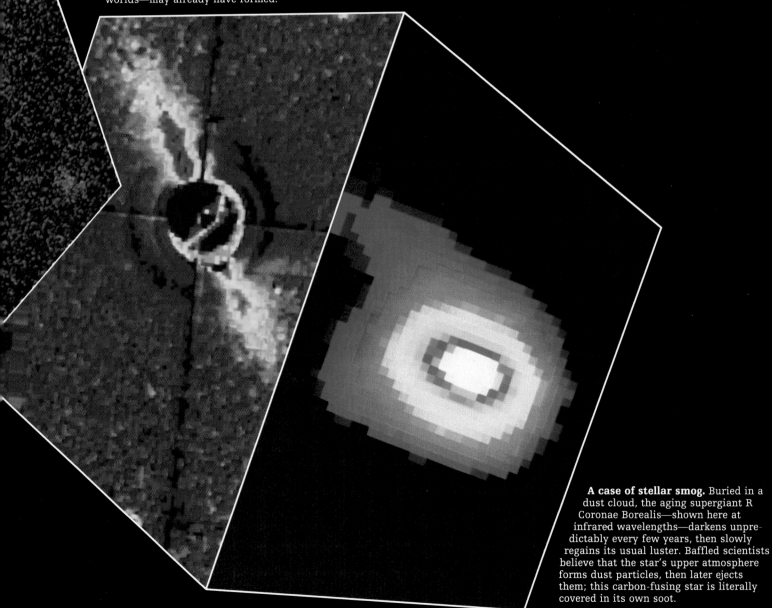

A case of stellar smog. Buried in a dust cloud, the aging supergiant R Coronae Borealis—shown here at infrared wavelengths—darkens unpredictably every few years, then slowly regains its usual luster. Baffled scientists believe that the star's upper atmosphere forms dust particles, then later ejects them; this carbon-fusing star is literally covered in its own soot.

first opportunity to make leisurely observations of celestial x-ray sources.

Most intriguing of all the x-ray objects known at the time of *Uhuru*'s launch was Cygnus X-1. Discovered by a rocket probe in 1962 and observed again on another rocket flight two years later, Cyg X-1, as it came to be known, was rated one of the most intense x-ray sources ever found. In 1971 astronomers realized that Cyg X-1 had several odd characteristics. For one, the intensity of its x-ray emissions varied rapidly. Since radiation from an object cannot vary any faster than the time it takes light to cross the object's surface, scientists knew that the source had to be remarkably small—smaller, in fact, than the Earth. Cyg X-1 also produced occasional radio signals. Researchers followed these signals to a bright blue supergiant star an estimated 6,500 light-years from Earth. By studying the Doppler shift in its spectra, they learned that the star was rapidly orbiting an unseen companion.

After estimating the mass of the blue supergiant (at least twenty solar masses) and studying the speed of its orbit, scientists deduced the mass of the unseen companion—at least six solar masses, twice the upper mass limit for a neutron star. The conclusion seemed inescapable: In all probability, Cyg X-1 was a black hole. Its x-rays might come from an accretion disk, a spiraling stream of gas drawn from the supergiant companion into an intensely hot whirlpool around the hole *(pages 123-135)*.

No one followed the course of the Cyg X-1 investigations more closely than Stephen Hawking, who placed a bet on the outcome with his friend and fellow relativity theorist Kip Thorne of Caltech. If Cyg X-1 was proved beyond a reasonable doubt to be a black hole, Hawking said, he would buy Thorne a one-year subscription to *Penthouse* magazine. If it turned out to be something other than a black hole, Hawking would win a one-year subscription to the British humor magazine *Private Eye.* (Hawking bet against Cyg X-1 for the sake of argument rather than conviction; both he and Thorne wanted and expected to see the black hole model confirmed.) So far, the bet remains unsettled, and Cyg X-1 is still only a strong suspect for black hole honors. But as *Uhuru* and subsequent x-ray satellites have accumulated data, more black hole candidates have appeared. Notable among them are the massive x-ray sources A0620-00, 3,000 light-years away in the constellation Monoceros, and LMC X-3, some 170,000 light-years distant in the Large Magellanic Cloud.

Despite supporting evidence, the existence of neutron stars and black holes depends heavily on the validity of Einstein's theories. In fact, the whole picture of late stellar evolution, not to mention cosmology, will have to be redrawn if relativity does not hold up. Perhaps the best words on the continuing struggle to understand the stars are those of Arthur Eddington. "Science is not just a catalogue of ascertained facts about the universe," he wrote in a 1927 book on stars. "It is a mode of progress, sometimes tortuous, sometimes uncertain. And our interest in science is not merely a desire to hear the latest facts added to the collection; we like to discuss our hopes and fears, probabilities and expectations. I have told the detective story as far as it has unrolled itself. I do not know whether we have reached the last chapter."

DENIZENS OF DARKNESS

The cataclysmic star death known as a supernova *(pages 89-101)* can produce two sorts of relics. One of these is a neutron star, an object perhaps ten miles in diameter but containing a million times the mass of Earth. The second kind of remnant—spawned by the explosion of a stellar giant at least ten times the mass of the Sun—is a so-called black hole, a construct of pure gravity so powerful that everything, including light, is trapped inside it.

Theoretically, huge numbers of black holes should exist. In the long life of the universe, uncountable billions of stars must have followed the inexorable path of stellar evolution to this dark end. One astrophysicist suggests that black holes in the Milky Way galaxy alone may well outnumber the visible stars and might account for galactic rotation rates that do not otherwise jibe with the visible stellar mass. In fact, black holes born from stars may be only one example of these gravitational prodigies. The enormous forces of the Big Bang may have created "mini" black holes that are scattered through space. And supermassive black holes, built from the matter of millions of stars, may occupy the center of many galaxies, including the Milky Way.

Finding objects that are invisible requires considerable detective work, but as explained on the following pages, astronomers are hot on the trail of this cosmic prey. Meanwhile, as the physics of black holes becomes clearer, these unseeable star remnants seem more marvelous than ever.

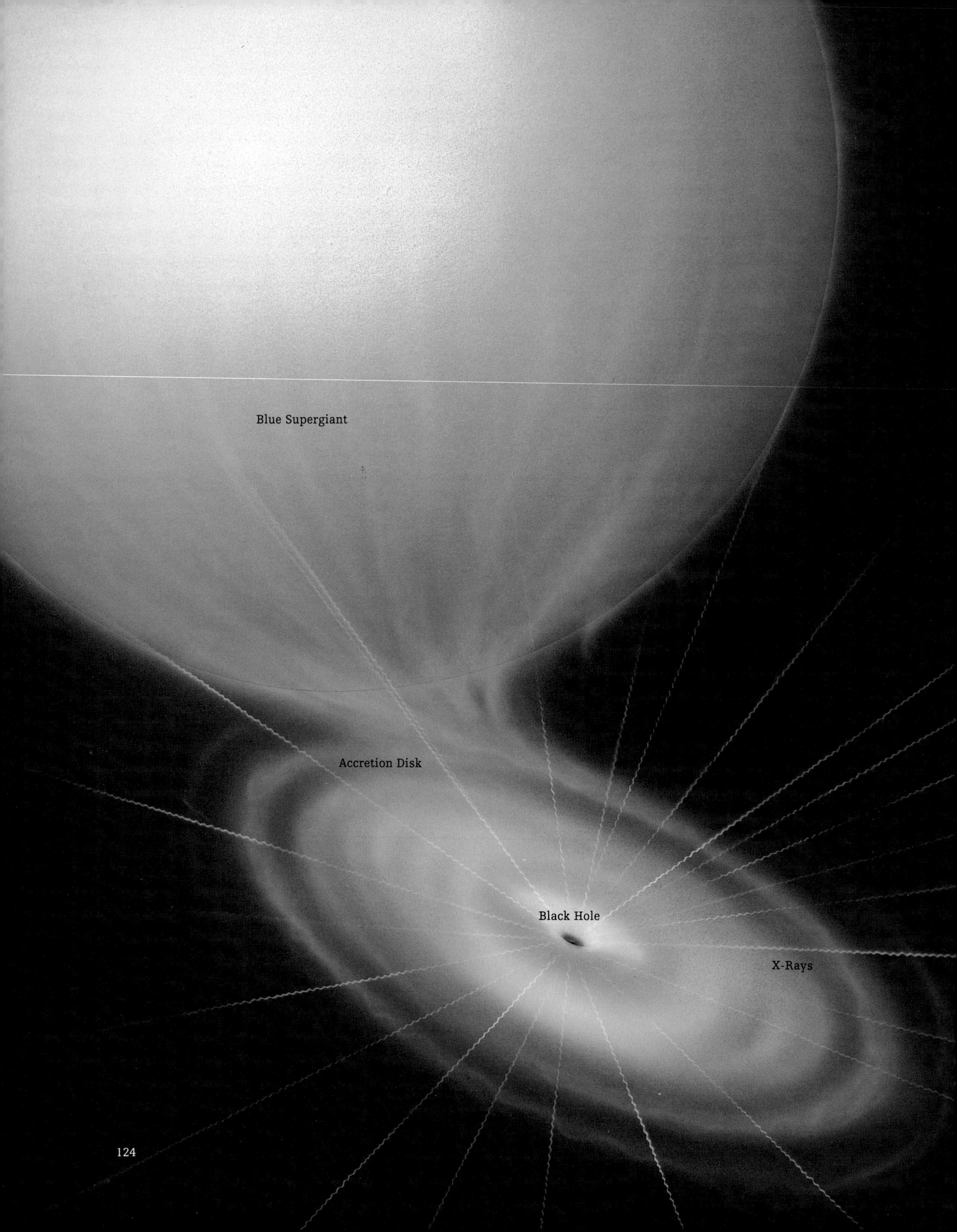

Blue Supergiant

Accretion Disk

Black Hole

X-Rays

Gravity offers some promising routes to the detection of stellar black holes. Some astronomers think the quarry will give itself away by emitting gravity waves, ripples in the fabric of space-time that are supposedly created when black holes form. Other researchers look for evidence of gravitational lensing: If a massive object—such as a black hole—is aligned properly between Earth and a distant star, distortions in space-time caused by the black hole will cause the star's light to bend, producing two identical star images. Gravitational lensing has been observed of very distant quasars, but these images do not necessarily require that a black hole be the lens; a massive but exceedingly faint galaxy would have the same effect.

So far, the best candidates for black hole status are so-called x-ray binaries—sources of x-rays that seem to be binary star systems. One such system, illustrated here, was discovered in 1971 in the constellation Cygnus, the Swan. The visible partner is a hot blue supergiant that is more than twenty times as massive as the Sun. Because blue supergiants do not emit x-rays, scientists assume the existence of an invisible source of the radiation, designated Cygnus X-1.

X-rays are produced by matter heated to several million degrees Fahrenheit. They could be generated by an accretion disk forming around a neutron star, but other observations eliminate that possibility in the case of Cygnus X-1. Spectral readings of the visible star show that it is engaged in an orbital dance, moving toward and away from Earth over a period of 5.6 days. Calculations based in part on the mass of the visible star and this rapid orbital period lead astronomers to predict that the unseen partner must contain the mass of at least six Suns, compressed into an area smaller than the Earth. Six solar masses constitute more than twice the estimated theoretical limit for neutron stars, leaving only one possible explanation: a black hole.

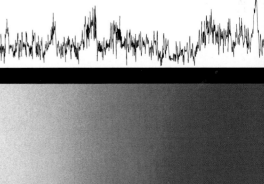

Matter pouring off a blue supergiant (left) forms an accretion disk about 2.5 million miles in diameter around Cygnus X-1, an invisible but extremely massive object believed to be a black hole. Because the inner edge of the disk orbits faster than the outer edge, friction heats the inner disk to more than three million degrees Fahrenheit, generating pulses of strong x-rays.

The chart at top records rapid x-ray pulses from Cygnus X-1, presumably generated by friction on the inner edge of an accretion disk rotating around a black hole. The optical spectrum (above) reflects the rotation of the visible star in the binary system. The shifting of the star's spectral lines toward the blue end of the spectrum and then toward the red indicates its approach to Earth and subsequent recession as it orbits an unseen companion.

1 The stage is set for the evolution of an x-ray binary system when two stars form close enough together to be bound in orbit by mutual gravity. In the sequence presented here—based on a model similar to Cygnus X-1 *(pages 118-119, 122)*—one star is significantly more massive than the other. Because of its immense size, this star fuses its hydrogen fuel at a much faster rate than its companion.

4 The supernova explosion blows much of the star's outer shell into space, but more than three solar masses of matter remain in the core. Crushed by its own gravity, this matter condenses to a singularity—a black hole from which even light cannot escape. The black hole continues to orbit its companion as before but is now invisible.

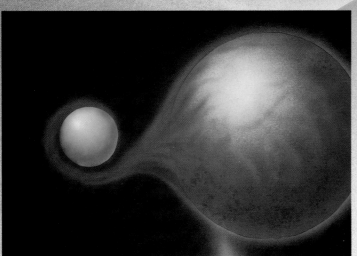

2 After a few million years, the more massive star begins to exhaust its store of hydrogen and shows the first signs of death: It enters the phase of stellar evolution known as the red giant phase, expanding to such an extent that gas from its outer shell streams across to the smaller star. The companion in turn begins to increase in size.

5 Because of its newly acquired extra mass, the companion star now begins a more rapid evolution, in the course of which it develops a stellar wind. This wind sweeps particles from the surface of the star toward the black hole, where they form a rapidly rotating accretion disk. Friction between the inner and outer edges of the disk generates temperatures high enough to cause the particles to emit intense x-rays—revealing the presence of the otherwise invisible black hole.

3 Having completely run out of hydrogen and other sources of fuel, the dying star suddenly explodes as a supernova. In a matter of seconds, its core begins a fateful and crushing collapse inward.

6 Within a few million years, only an eye blink on the stellar time scale, the companion star also evolves to the red giant phase. As it dumps increasing amounts of gas onto the black hole's accretion disk, the disk thickens, absorbing the x-rays generated at its inner edge until the black hole once again becomes invisible.

EVOLUTION OF AN X-RAY BINARY

The most successful method so far for finding black hole candidates is examining x-ray binary systems, but the method has certain limitations. Although binary stars number perhaps a third of all the stars in the Milky Way galaxy, only a very small fraction of such pairs will ultimately become x-ray binaries. For the necessary interactions to take place, one partner must begin life with a mass greater than ten Suns, and the pair must orbit each other closely enough for matter to be exchanged.

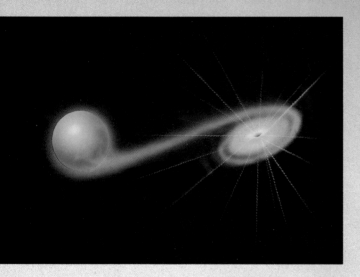

Even binary systems that do eventually produce x-rays might not harbor a black hole. The emission of x-rays indicates that matter is being pulled from the visible star onto an incredibly dense and compact object, but only by estimating the mass of the system's unseen player—from the mass and orbital speed of the visible partner—can astronomers determine whether they have in fact found the object of their search or just a neutron star.

Calculating the mass of the invisible half of the x-ray binary is not easy. It requires an accurate estimate of the mass of the visible partner—something scientists have long been able to do for solitary stars from measurements of their luminosity and color. But the accuracy of those measurements may be suspect for a star in close orbit with a companion.

Finally, even assuming all calculations are perfect, scientists face yet another problem: X-ray binaries emit this radiation for only about one-half of one percent of the few million years of their existence. Catching these cosmic fireflies requires as much serendipity as perseverance.

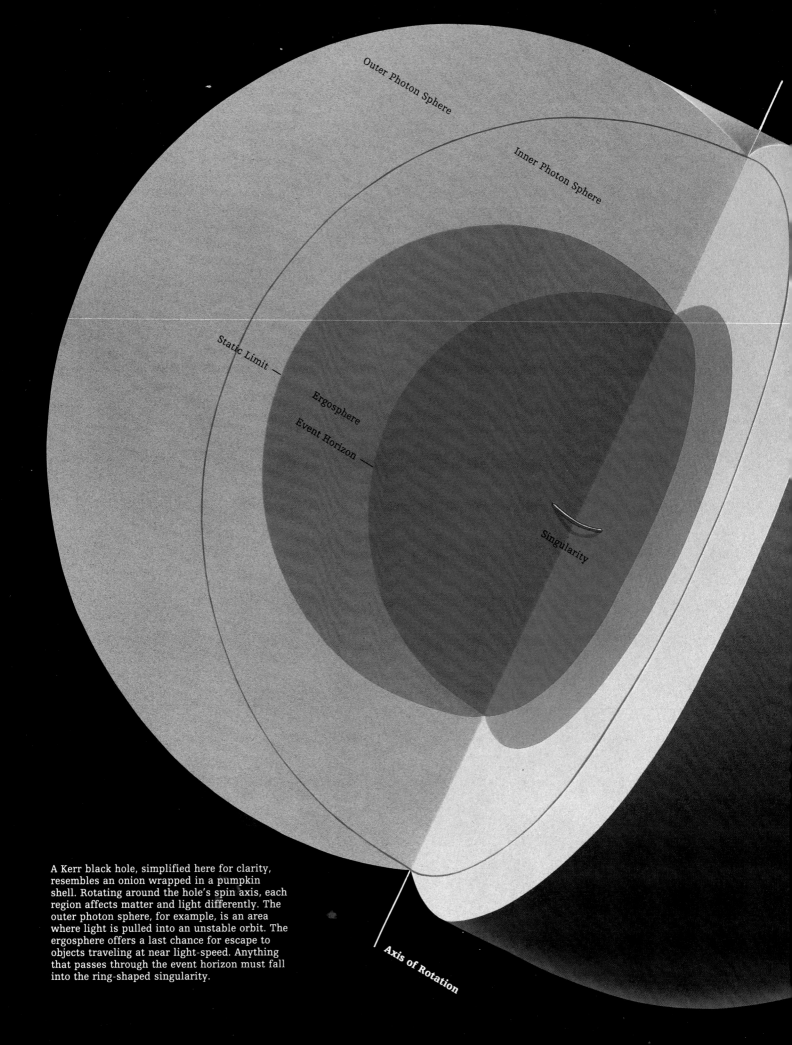

Outer Photon Sphere

Inner Photon Sphere

Static Limit —

Ergosphere

Event Horizon —

Singularity

Axis of Rotation

A Kerr black hole, simplified here for clarity, resembles an onion wrapped in a pumpkin shell. Rotating around the hole's spin axis, each region affects matter and light differently. The outer photon sphere, for example, is an area where light is pulled into an unstable orbit. The ergosphere offers a last chance for escape to objects traveling at near light-speed. Anything that passes through the event horizon must fall into the ring-shaped singularity.

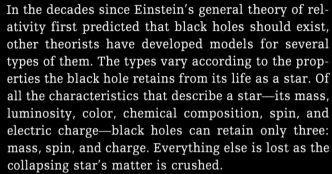

ARCHITECTURE OF THE INVISIBLE

In the decades since Einstein's general theory of relativity first predicted that black holes should exist, other theorists have developed models for several types of them. The types vary according to the properties the black hole retains from its life as a star. Of all the characteristics that describe a star—its mass, luminosity, color, chemical composition, spin, and electric charge—black holes can retain only three: mass, spin, and charge. Everything else is lost as the collapsing star's matter is crushed.

By definition, all black holes possess mass. In the simplest type, known as the Schwarzschild black hole for German astronomer Karl Schwarzschild, mass is the object's only property, and all of it is centered on a point of infinite density called a singularity. In other types, black holes are assumed to have various combinations of the three properties. In the real universe, however, a black hole with charge is an unlikely phenomenon, because a massive star with an excess of positive or negative charge would quickly be neutralized by the attraction of opposite charge.

Illustrated here and on the next six pages is the Kerr black hole, the type believed most likely to arise in the course of stellar evolution. Named after University of Texas scientist Roy Kerr, it possesses both mass and spin. The physics resulting from the hole's rotation around an axis produces a singularity that is not a point as in the Schwarzschild model but a ring. In addition, the hole's rotation pulls space-time itself around with it, a phenomenon known as frame dragging. The area surrounding the ring-shaped singularity is separated into domains of differing characteristics. The outer and inner photon spheres are regions where light hitting at the proper angles is pulled into orbit *(pages 130-131).* In a region called the ergosphere, marked by a boundary known as the static limit, frame dragging dictates that an object cannot remain at rest because space-time itself is in motion around the singularity. Within the ergosphere, it is still theoretically possible to escape from the black hole, but once an object passes through the next boundary—the event horizon—escape is no longer an option. Beyond that surface, unimaginable gravity overpowers everything, even light.

At the surface of the outer photon sphere of a rotating black hole, light that normally radiates equally in all directions begins to be affected by immense gravity. Light rays shining at angles close to the surface either get pulled into an orbit within the photon sphere or fall into the black hole.

If a light source descends halfway into the photon sphere, only light rays traveling nearly perpendicular to the surface can escape; rays traveling at other angles are overpowered by gravity. Light escaping through the narrowing exit cone is distorted by the black hole's spin.

Just above the event horizon, gravity squeezes the exit cone to pencil thinness: Only light rays traveling absolutely perpendicular to the surface can escape—and then only by following a circuitous route dictated by the black hole's gravity and rotation.

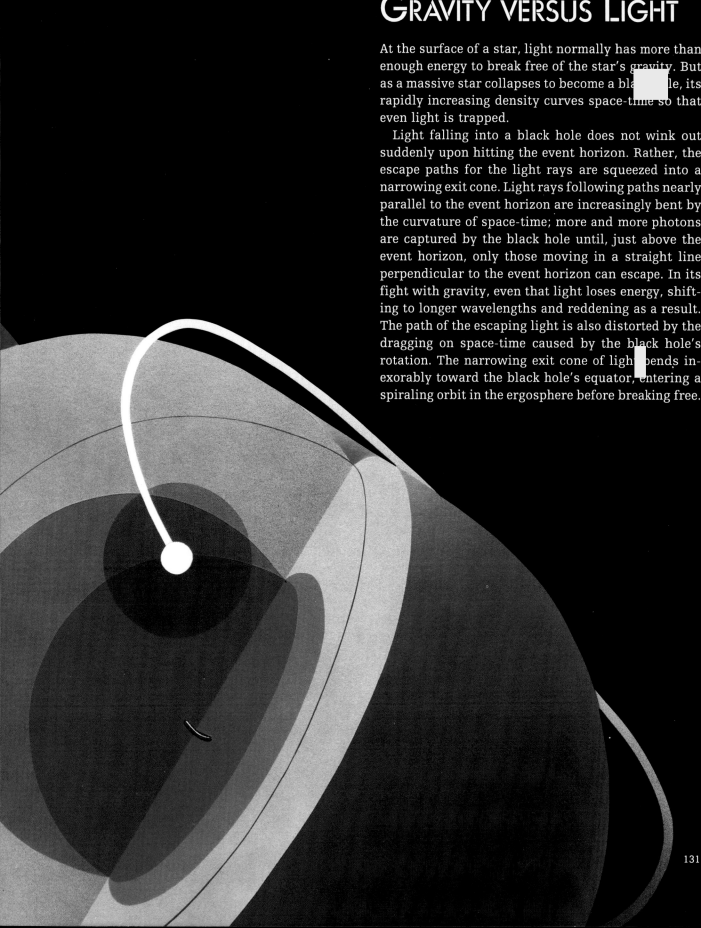

GRAVITY VERSUS LIGHT

At the surface of a star, light normally has more than enough energy to break free of the star's gravity. But as a massive star collapses to become a black hole, its rapidly increasing density curves space-time so that even light is trapped.

Light falling into a black hole does not wink out suddenly upon hitting the event horizon. Rather, the escape paths for the light rays are squeezed into a narrowing exit cone. Light rays following paths nearly parallel to the event horizon are increasingly bent by the curvature of space-time; more and more photons are captured by the black hole until, just above the event horizon, only those moving in a straight line perpendicular to the event horizon can escape. In its fight with gravity, even that light loses energy, shifting to longer wavelengths and reddening as a result. The path of the escaping light is also distorted by the dragging on space-time caused by the black hole's rotation. The narrowing exit cone of light bends inexorably toward the black hole's equator, entering a spiraling orbit in the ergosphere before breaking free.

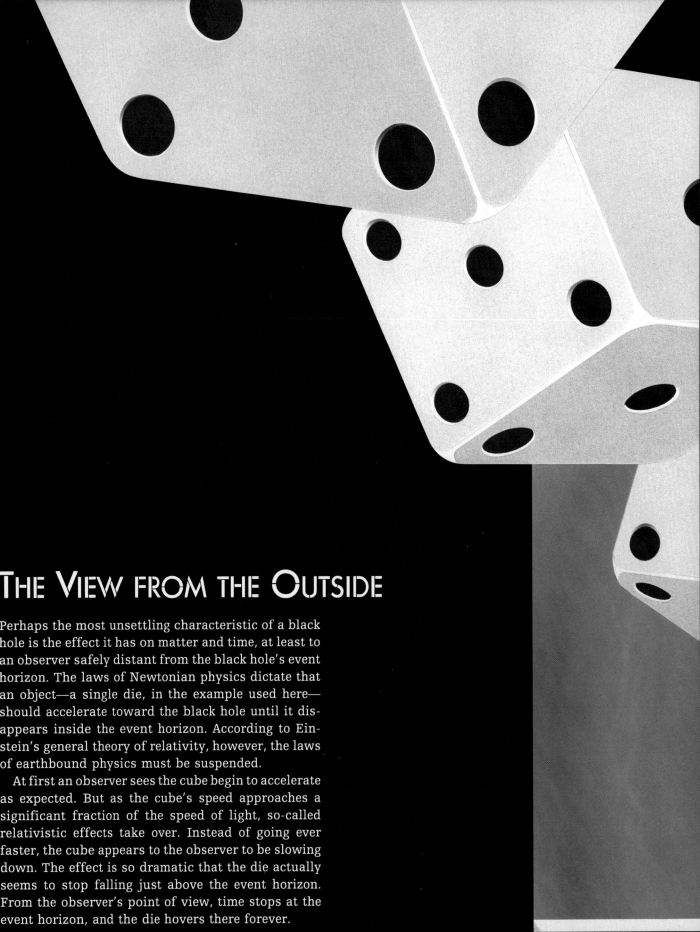

The View from the Outside

Perhaps the most unsettling characteristic of a black hole is the effect it has on matter and time, at least to an observer safely distant from the black hole's event horizon. The laws of Newtonian physics dictate that an object—a single die, in the example used here—should accelerate toward the black hole until it disappears inside the event horizon. According to Einstein's general theory of relativity, however, the laws of earthbound physics must be suspended.

At first an observer sees the cube begin to accelerate as expected. But as the cube's speed approaches a significant fraction of the speed of light, so-called relativistic effects take over. Instead of going ever faster, the cube appears to the observer to be slowing down. The effect is so dramatic that the die actually seems to stop falling just above the event horizon. From the observer's point of view, time stops at the event horizon, and the die hovers there forever.

The path of a die falling toward a black hole initially follows the laws of conventional Newtonian physics: It accelerates and falls in a spiraling path. But to a distant observer, the cube then appears to slow down just as it nears the speed of light. In addition, the increasing gravity causes light reflected from the die to lengthen and red-shift.

As a die begins its fall toward a black hole in the first frame of this imaginary filmstrip, the event horizon —the surface from which not even light can escape—appears as a small black circle. The event horizon looms ever larger with the passage of time as the die accelerates toward the speed of light.

By the third frame of the filmstrip, the die is only a few miles above the event horizon and suffering the effects of the warping of space-time. Stretched almost beyond recognition, the cube begins to show signs of fracturing. The photon sphere, filled with trapped, orbiting light, appears as a circle around the lightless event horizon.

In the final frame, the die's distress in the highly warped fabric of space-time is so severe that the cube begins to break apart, its leading corner first.

The grossly elongated pieces of the die, now through the surface of the event horizon, continue to stretch and accelerate to near the speed of light. To the theoretical observer accompanying these pieces, starlight that is entering the region appears blue-shifted because it is gaining energy as it approaches the yawning blackness hiding the infinitely dense singularity.

RACING TOWARD DESTRUCTION

If to an outsider the falling die never reaches its destination, the perception of a theoretical observer perched on one corner of the cube is drastically different. Because the observer is in the same frame of space-time as the die, time runs normally and the cube accelerates toward the event horizon as would be expected even by Newton's laws. But as the event horizon nears, space-time is powerfully stretched by the rapidly increasing gravity of the black hole. Embedded in this mangled fabric, the hapless cube is stretched from front to back and squeezed from the sides. Long before it reaches the event horizon, the die is ripped apart by the powerful tidal forces.

Fragments of the die continue to be stretched as they fall into the black hole, but the observer would perceive no sudden change to mark the crossing of the event horizon. As the fragments near the speed of light, stars in the outside universe appear increasingly distorted and their images tend to concentrate in front of the die.

GLOSSARY

Absorption line: a dark line or band at a particular wavelength on a spectrum, formed when a substance between a radiating source and an observer absorbs electromagnetic radiation of that wavelength. Different substances produce characteristic patterns of absorption lines.

Accretion disk: a disk formed from gases and other materials drawn in by a compact body, such as a black hole or a neutron star, at the disk's center.

Alpha particle: a helium nucleus, consisting of two protons and two neutrons. In the so-called triple-alpha process, three alpha particles fuse to form a carbon nucleus.

Angstrom: a unit of length equal to one ten-billionth of a meter (about four-billionths of an inch); used in astronomy as a measure of wavelength.

Angular momentum: a measure of an object's inertia, or state of motion, about an axis of rotation.

Atmosphere: the gaseous outer layer of a planet, star, or other body with sufficient gravity to maintain it.

Atom: the smallest component of a chemical element that retains the properties associated with that element. Atoms are composed of protons, neutrons, and electrons; the number of protons determines the identity of the element.

Binary star system: a gravitationally bound pair of stars in orbit around their mutual center of gravity. Binary stars are extremely common, as are systems of three or more stars.

Black dwarf: the hypothetical remnant of a dwarf star that has completely consumed its nuclear fuel.

Black hole: theoretically, an extremely compact body with such great gravitational force that no radiation can escape from it. Proposed varieties include mini black holes, low-mass objects from the beginning of the universe; stellar black holes, which form from the cores of old, very massive stars; and supermassive black holes, equivalent to several million stars in mass and located in the centers of galaxies.

Blue giant: a young, luminous, high-mass star of spectral class O or B.

Brown dwarf: a dim body of less than 0.1 solar mass with not enough self-gravity to fuse hydrogen to helium.

Carbon cycle: a fusion reaction that occurs in high-mass stars, in which carbon, oxygen, and nitrogen nuclei provide sites for hydrogen nuclei to fuse to helium.

Celestial coordinates: a pair of numbers designating an object's location on the celestial sphere. One coordinate, declination, is a north-south value similar to latitude; the other, right ascension, is similar to longitude.

Celestial sphere: the apparent sphere of sky that surrounds the Earth; used by astronomers as a convention for specifying the location of a celestial object.

Celsius: a scientific temperature scale in which 0 degrees is the freezing point and 100 degrees the boiling point of water.

Chandrasekhar limit: the maximum core mass for a stable white dwarf star, about 1.44 solar masses; named for Subrahmanyan Chandrasekhar, who proposed it. Stars that exceed the limit may become neutron stars or black holes.

Charge-coupled device (CCD): an electronic array of detectors, usually positioned at a telescope's focus, for registering electromagnetic radiation.

Constellation: originally a pattern of stars named for an object, animal, or person but now more commonly the area of sky assigned to that pattern. Every astronomical object is located in a specific constellation.

Continuous spectrum: a spectrum consisting of all wavelengths in a given range, without absorption or emission lines.

Convection: the transfer of heat in a fluid or gas by the movement of currents from hotter to cooler regions.

Core: the central component of a celestial body. In stars, the core is where fusion usually occurs.

Cosmic ray: an atomic nucleus or other charged particle moving at close to the speed of light.

Cosmology: the study of the universe as a whole, including its large-scale structure and movements, its origin, and its ultimate fate.

Degeneracy: a highly compressed state of matter characteristic of the core of certain dwarf stars. In degenerate matter, the ideal gas law relating density to pressure does not apply.

Density: the ratio of mass to volume in an object. A white dwarf, for example, has a density of one hundred thousand trillion kilograms per cubic meter.

Deuteron: a particle consisting of one neutron and one proton; equivalent to the nucleus of an atom of deuterium, an isotope of hydrogen.

Doppler effect: a wave phenomenon in which waves appear to compress as their source approaches the observer or stretch out as the source recedes from the observer.

Electromagnetic radiation: radiation consisting of periodically varying electric and magnetic fields that vibrate perpendicularly to each other and travel through space at the speed of light.

Electromagnetic spectrum: the array, in order of frequency or wavelength, of electromagnetic radiation, from low-frequency, long-wavelength radio waves, through infrared, visible light, and ultraviolet, to high-frequency, short-wavelength gamma rays.

Electron: a negatively charged particle that normally orbits an atom's nucleus but may exist in isolation.

Emission line: a bright band at a particular wavelength on a spectrum, emitted directly by the source and indicating by its wavelength a chemical constituent of that source.

Energy level: the quantity of energy associated with an electron. An increase in energy will shift electrons to higher energy levels within an atom.

Ergosphere: the region around a black hole, inside the photon sphere and outside the event horizon, where only objects that remain in motion can avoid entering the singularity.

Event horizon: the boundary around a black hole's singularity, within which gravitational forces prevent anything, including light, from escaping.

Exclusion principle: the rule of quantum mechanics stating that no two electrons, neutrons, or protons with the same energy, angular momentum, and spin can exist simultaneously in the same atom. Also called the Pauli exclusion principle after its formulator, Wolfgang Pauli.

Fahrenheit: a nonastronomical temperature scale in which 32 degrees is equivalent to 0 degrees Celsius and 212 degrees is equivalent to 100 degrees Celsius.

Frequency: the number of oscillations per second of an electromagnetic (or other) wave. *See* wavelength.

Fusion: the combining of two atomic nuclei to form a heavier nucleus, releasing great energy as a by-product.

Gamma rays: the most energetic form of electromagnetic radiation, with the highest frequency and shortest wavelength of all forms on the spectrum.

Gas: a fluid state of matter in which atoms and molecules are not bound together and will diffuse unless contained; may also refer to an interstellar material composed primarily of hydrogen and helium gases.

Giant molecular cloud: a concentration of interstellar gas and dust up to several dozen light-years in diameter.

Gravity: the mutual attraction of separate masses; a fundamental force of nature.

Gravity wave: a theoretical perturbation in an object's gravitational field that would travel at the speed of light. Relativity theory predicts that gravity waves may result from accelerating, oscillating, or violently disturbed masses—categories that include black holes, neutron stars, and supernovae.

Ground state: the lowest possible energy level for a given electron.

Helium: the second lightest chemical element and the second most abundant; produced in stars by the fusion of hydrogen.

Helium flash: a brief stage in the evolution of stars with a mass of approximately two and a half solar masses or less, in which the fusion of degenerate helium to carbon spreads in a "flash" throughout the core.

Hertzsprung-Russell diagram (H-R diagram): a graph of stellar properties that charts absolute magnitude against spectral class.

Hydrogen: the most common detectable element in the universe. Stellar energy comes primarily from the fusion of hydrogen nuclei.

Ideal gas law: the principle that for a given quantity of gas under normal conditions, the gas's pressure is directly proportional to the product of its density and temperature.

Infrared: a band of electromagnetic radiation with a lower frequency and a longer wavelength than visible red light.

Intensity: the amount of radiation received from an object; optical astronomers prefer the term *brightness*.

Inverse-square law: the mathematical description of how some forces, including electromagnetism and gravity, weaken in inverse proportion to the square of the distance from the source.

Ion: an atom that has lost or gained one or more electrons. In comparison, the neutral atom has an equal number of electrons and protons, giving it a zero net electrical charge. A positive ion has fewer electrons than the neutral atom, a negative ion has more.

Isotope: one of two or more forms of a chemical element that have the same number of protons but a different number of neutrons in the nucleus.

Kelvin: an absolute temperature scale that uses Celsius degrees but sets 0 degrees at absolute zero, about minus 273 degrees Celsius; named for British physicist William Thomson, Lord Kelvin.

Light-year: an astronomical distance unit equal to the distance light travels in a vacuum in one year, almost six trillion miles.

Luminosity: an object's total energy output, usually measured in ergs per second.

Magnetic pole: one of two locations on the surface of a neutron star or other body where its magnetic field lines converge, as opposed to the poles defined by the body's axis of rotation.

Magnitude: a designation of an object's brightness or luminosity relative to that of other objects; *apparent magnitude* refers to observed brightness, *absolute magnitude* to an object's hypothetical brightness at a standard distance of about 32.6 light-years. By convention, absolute magnitude is used in astronomy to indicate an object's actual luminosity.

Main sequence: a diagonal region in the Hertzsprung-Russell diagram that includes 90 percent of all stars.

Mass: a measure of the total amount of material in an object, determined either by its gravity or by its tendency to stay in motion, if in motion, or at rest, if at rest.

Molecule: the smallest particle of an element or compound that retains its properties. A molecule consists of one or more atoms bonded together.

Nebula: a cloud of interstellar dust or gas; in some cases a supernova remnant or a shell ejected by a star. The term nebula once included all soft-edged objects, such as galaxies and globular clusters; this usage is no longer correct.

Neutrino: a chargeless particle with little or no mass that moves at close to the speed of light.

Neutron: an uncharged particle with a mass similar to that of a proton; normally found in an atom's nucleus.

Neutron star: a very dense body composed of tightly packed neutrons; one possible product of a supernova explosion. Neutron stars are observed as pulsars.

Noise: meaningless random changes in radiation that tend to obscure a specific signal.

Nova: a star that exhibits a sudden, temporary increase in brightness thousands of times its normal appearance.

Nuclear fusion: *see* Fusion.

Nucleus: the massive center of an atom, composed of protons and neutrons and orbited by electrons.

Parallax: a star's apparent motion on the celestial sphere over a six-month period. Measured in seconds of arc, it is used to determine a star's distance; the greater the parallax, the nearer the star.

Parsec: an astronomical distance unit equivalent to about 3.26 light-years. A star would be one parsec from Earth if it exhibited a *parallax* of one *second* of arc.

Photometer: a device that measures an object's brightness, or apparent magnitude, by detecting its emitted photons.

Photon: a unit of electromagnetic energy associated with a specific wavelength.

Photon sphere: a region around a black hole that captures light traveling at particular angles; bounded on the inside by the static limit.

Plasma: a gaslike association of ionized particles that responds collectively to electric and magnetic fields. Because plasma particles do not interact the way particles of ordinary gas do, plasma is considered a fourth state of matter, along with solid, liquid, and gas.

Positron: a particle similar to an electron but positively charged.

Proton: a positively charged particle with about 2,000 times the mass of an electron; normally found in an atom's nucleus.

Proton-proton reaction: a fusion reaction that predominates in stars of two and a half solar masses or less, in which hydrogen nuclei, consisting of single protons, fuse to form deuterium and, ultimately, helium.

Protostar: a large gaseous sphere, held together by its own gravitational attraction, that shrinks and compresses to become a star.

Pulsar: a radiating source that emits extremely regular

bursts of energy at intervals of several seconds or less. Pulsars are almost certainly neutron stars.

Radio: the least energetic form of electromagnetic radiation, with the lowest frequency and the longest wavelength.

Red dwarf: a dim, long-lived, low-mass, M-class star.

Red giant: an aging, low-mass star that has greatly expanded and cooled after consuming most of its core hydrogen; usually of spectral class M.

Relativity: a set of related theories in modern physics that show among other effects that mass and energy are equivalent and that mass, geometry, and time are measured differently by observers in relative motion or in varying gravitational fields.

Resolution: the degree to which details in an image can be separated, or resolved. The resolving power of a telescope is usually proportional to the diameter of its mirror.

Singularity: the infinitely condensed mass at the center of a black hole that has no dimensions in the physical universe.

Solar mass: a stellar mass unit equal to the Sun's mass, about two thousand trillion trillion grams.

Spectral class: a star's classification, based on its spectrum, according to an established system.

Spectrogram: a photograph of an astronomical spectrum.

Spectrograph: an instrument that splits light or other electromagnetic radiation into its individual wavelengths, or spectrum, and records the result photographically.

Spectroscopy: the study of spectra, including the position and intensity of emission and absorption lines.

Spectrum: the array of colors or frequencies obtained by dispersing light, as through a prism; often banded with absorption or emission lines.

Spin: a mathematical property of subatomic particles that is analogous to the angular momentum of a spinning top. Spin can be either positive or negative.

Static limit: the outer boundary of the ergosphere of a black hole, within which objects must remain in motion to avoid entering the singularity.

Supergiant: an old, high-mass star greatly expanded from its original size; larger and brighter than a giant star.

Supernova: a stellar explosion that expels all or most of the star's mass and is extremely luminous.

Supernova remnant: an expanding nebula, consisting of the stellar matter ejected by a supernova.

Triple-alpha process: a fusion reaction, characteristic of red giants and other highly evolved stars, in which three helium nuclei, also called alpha particles, fuse to carbon.

T Tauri object: a very young star characterized by extensive and violent ejections of its mass; named after the first known star of this type.

Ultraviolet: a band of electromagnetic radiation with a higher frequency and shorter wavelength than visible blue light.

Velocity: the speed and direction of motion.

Wave: the propagation of a pattern of disturbance through a material medium.

Wavelength: the distance from crest to crest or trough to trough of an electromagnetic or other wave. Wavelengths are related to frequency: the longer the wavelength, the lower the frequency.

White dwarf: an old, extremely dense star about as large as the Earth but with a mass as great as the Sun's; the remains of a star that has completely fused its helium core.

X-rays: a band of electromagnetic radiation intermediate in wavelength between ultraviolet radiation and gamma rays.

BIBLIOGRAPHY

Books

Abbott, David, ed., *Astronomers.* New York: Peter Bedrick Books, 1984.

Abell, George O., David Morrison, and Sidney C. Wolff, *Exploration of the Universe.* Philadelphia: Saunders College Publishing, 1987.

Abetti, Giorgio, *The History of Astronomy.* Transl. by Betty Burr Abetti. New York: Henry Schuman, 1952.

Alter, Dinsmore, Clarence H. Cleminshaw, and John G. Phillips, *Astronomy.* New York: Thomas Y. Crowell, 1969.

Asimov, Isaac, *Asimov's Biographical Encyclopedia of Science and Technology.* New York: Doubleday, 1982.

Bartusiak, Marcia, *Thursday's Universe.* New York: Times Books, 1986.

Berger, Melvin, *Bright Stars, Red Giants and White Dwarfs.* New York: G. P. Putnam's Sons, 1983.

Berman, Louis, and J. C. Evans, *Exploring the Cosmos.* Boston: Little, Brown, 1986.

Bernstein, Jeremy, *Hans Bethe: Prophet of Energy.* New York: Basic Books, 1980.

Berry, Richard, *Discover the Stars.* New York: Harmony Books, 1987.

Black, David C., and Mildred Shapley Matthews, eds., *Protostars & Planets II.* Tucson: University of Arizona Press, 1985.

Boslough, John, *Stephen Hawking's Universe.* New York: William Morrow, 1985.

Broad, William, and Nicholas Wade, *Betrayers of the Truth.* New York: Simon & Schuster, 1982.

Burnham, Robert, Jr., *Burnham's Celestial Handbook.* 3 vols. New York: Dover, 1978.

Clark, David H., *Superstars.* New York: McGraw-Hill, 1984.

Cooke, Donald A., *The Life & Death of Stars.* New York: Crown, 1985.

Crowther, James Gerald, *Six Great Astronomers.* London: Hamish Hamilton Ltd., 1961.

Feldman, Anthony, *Space.* New York: Facts on File, 1980.

Fredrick, Laurence W., and Robert H. Baker, *An Introduction to Astronomy.* New York: D. Van Nostrand, 1974.

Friedman, Herbert, *The Amazing Universe.* Washington, D.C.: National Geographic Society, 1975.

Gamow, George, *A Star Called the Sun.* New York: Viking, 1964.

Gingerich, Owen, ed., *The General History of Astronomy: Astrophysics and Twentieth-Century Astronomy to 1950: Part A* (Vol. 4). New York: Cambridge University Press, 1984.

Greenstein, George, *Frozen Star.* New York: Freundlich Books, 1983.

Haramundanis, Katherine, ed., *Cecilia Payne-Gaposchkin:*

An Autobiography and Other Recollections. New York: Cambridge University Press, 1984.

Hartmann, William K., *Astronomy: The Cosmic Journey.* Belmont, Calif.: Wadsworth, 1987.

Hawking, S. W., and W. Israel, eds., *Three Hundred Years of Gravitation.* New York: Cambridge University Press, 1987.

Hearnshaw, J. B., *The Analysis of Starlight.* New York: Cambridge University Press, 1986.

Hoyle, Fred, *The Physics-Astronomy Frontier.* New York: W. H. Freeman, 1980.

Jastrow, Robert, and Malcolm H. Thompson, *Astronomy: Fundamentals and Frontiers.* New York: John Wiley & Sons, 1974.

Jones, Bessie Zaban, and Lyle Gifford Boyd, *The Harvard College Observatory: The First Four Directorships, 1839-1919.* Cambridge, Mass.: Harvard University Press, 1971.

Kaler, James B., *Stars and Their Spectra: An Introduction to the Spectral Sequence.* Cambridge, Mass.: Cambridge University Press, 1989 (in press).

Kaufmann, William J., III:
The Cosmic Frontiers of General Relativity. Boston: Little, Brown, 1977.
Universe. New York: W. H. Freeman, 1985.

Kippenhahn, Rudolf, *100 Billion Suns: The Birth, Life, and Death of the Stars.* Transl. by Jean Steinberg. New York: Basic Books, 1983.

Lampton, Christopher, *Black Holes and Other Secrets of the Universe.* New York: Franklin Watts, 1980.

Lang, Kenneth R., and Owen Gingerich, eds., *A Source Book in Astronomy and Astrophysics, 1900-1975.* New York: Cambridge University Press, 1984.

Malin, David, and Paul Murdin, *Colours of the Stars.* New York: Cambridge University Press, 1984.

Motz, Lloyd, and Anneta Duveen, *Essentials of Astronomy.* New York: Columbia University Press, 1977.

Murdin, Paul, and David Allen, *Catalogue of the Universe.* New York: Cambridge University Press, 1979.

Murdin, Paul, and Lesley Murdin, *Supernovae.* New York: Cambridge University Press, 1985.

Page, Thornton, and Lou Williams Page, *Starlight: What It Tells about the Stars* (Vol. 5). New York: Macmillan, 1967.

Pasachoff, Jay M., *Contemporary Astronomy.* Philadelphia: W. B. Saunders, 1977.

Payne-Gaposchkin, Cecilia, *Stars in the Making.* Cambridge, Mass.: Harvard University Press, 1961.

Preston, Richard, *First Light: The Search for the Edge of the Universe.* New York: Atlantic Monthly Press, 1987.

Ronan, Colin A., *Astronomers Royal.* Garden City, N.Y.: Doubleday, 1969.

Shapley, Harlow, ed., *Source Book in Astronomy: 1900-1950.* Cambridge, Mass.: Harvard University Press, 1960.

Shipman, Harry L., *Black Holes, Quasars, and the Universe.* Boston: Houghton Mifflin, 1980.

Smith, F. G., *Pulsars.* New York: Cambridge University Press, 1977.

Snow, Theodore P., *Essentials of the Dynamic Universe.* St. Paul, Minn.: West, 1987.

Struve, Otto, and Velta Zebergs, *Astronomy of the 20th Century.* New York: Macmillan, 1962.

Sullivan, Walter, *Black Holes: The Edge of Space, the End of Time.* Garden City, N.Y.: Anchor Press/Doubleday, 1979.

Tucker, Wallace, and Riccardo Giacconi, *The X-Ray Universe.* Cambridge, Mass.: Harvard University Press, 1985.

Vehrenberg, Hans, *Atlas of Deep-Sky Splendors.* Cambridge, Mass.: Sky, 1978.

Verschuur, Gerrit L., *The Invisible Universe Revealed.* New York: Springer-Verlag, 1987.

Wyatt, Stanley P., and James B. Kaler, *Principles of Astronomy.* Boston: Allyn and Bacon, 1981.

Periodicals

Allen, David A., "Star Formation and IRAS Galaxies." *Sky & Telescope,* April 1987.

Andersen, Per H., "Mystery Spots, X Rays, Gamma Rays: Is the Dust Settling from SN1987a?" *Physics Today,* January 1988.

Bartusiak, Marcia, "Celestial Zoo." *Omni,* December 1982.

"Black Holes: Stopping Time, Powering Galaxies." *National Geographic,* June 1983.

Boslough, John, "The Unfettered Mind." *Science,* November 1981.

Burbidge, E. Margaret, et al., "Synthesis of the Elements in Stars." *Reviews of Modern Physics,* October 1957.

Burbidge, Geoffrey, David Layzer, and John G. Phillips, eds., *Annual Review of Astronomy and Astrophysics,* Vol. 24. Palo Alto, Calif.: Annual Reviews Inc., 1986.

Cohen, Martin, "Do Supernovae Trigger Star Formation?" *Astronomy,* April 1982.

Croswell, Ken, "Stars Too Small to Burn." *Astronomy,* April 1984.

"Echoing Supernova." *Science News,* December 5, 1987.

"Fantastic Signals from Space." *Time,* March 15, 1968.

Helfand, David, "Bang: The Supernova of 1987." *Physics Today,* August 1987.

Hoyle, Fred, "The Universe: Past and Present Reflections." *Annual Review of Astronomy and Astrophysics,* Vol. 20. Palo Alto, Calif.: Annual Reviews Inc., 1982.

Hutchings, John B., and David Crampton, "LMC X-3: A Black Hole." *Mercury,* July/August 1984.

Kaler, James B., "Origins of the Spectral Sequence." *Sky & Telescope,* February 1986.

Kawaler, Steven D., and Donald E. Winget, "White Dwarfs: Fossil Stars." *Sky & Telescope,* August 1987.

Kirshner, Robert P., "Supernova: Death of a Star." *National Geographic,* May 1988.

McClintock, Jeffrey, "Stalking the Black Hole in the Star Garden of the Unicorn." *Mercury,* July/August 1987.

Maran, Stephen P., "A Blue Supergiant Dissects Itself in a Cosmic Explosion." *Smithsonian,* April 1988.

Overbye, Dennis:
"Exploring the Edge of the Universe." *Discover,* December 1982.
"The Man Who Believes in Forever." *Discover,* May 1981.
"Out from under the Cosmic Censor: Stephen Hawking's Black Hole." *Sky & Telescope,* August 1977.

Penrose, Roger, "Black Holes." *Scientific American,* May 1972.

Rothschild, R. E., et al., "Millisecond Temporal Structure in CYG X-1." *Astrophysical Journal,* 1974, p. L13.

Rubin, Vera, "Women's Work." *Science 86,* July/August 1986.

Schorn, Ronald A.:
"Binary Pulsars: Back from the Grave." *Sky & Telescope,* December 1986.
"Happy Birthday, Supernova!" *Sky & Telescope,* February 1988.

Schwartzenburg, Dewey, "Brightness and Magnitude." *Astronomy,* July 1979.

Scoville, Nick, and Judith S. Young, "Molecular Clouds, Star Formation and Galactic Structure." *Scientific American,* April 1984.

Seward, Frederick D., Paul Gorenstein, and Wallace H. Tucker, "Young Supernova Remnants." *Scientific American,* August 1985.

Shu, Frank H., Fred C. Adams, and Susana Lizano, "Star Formation in Molecular Clouds: Observation and Theory." *Annual Review of Astronomy and Astrophysics,* Vol. 25. Palo Alto, Calif.: Annual Reviews Inc., 1987.

Talcott, Richard, "Insight into Star Death." *Astronomy,* February 1988.

"Third Black Hole Discovered." *Astronomy,* June 1986.

Thorne, Kip S., "The Search for Black Holes." *Scientific American,* December 1974.

Tierney, John, "Exploding Star Contains Atoms of Elvis Presley's Brain." *Discover,* July 1987.

Trimble, Virginia, "A Field Guide to Close Binary Stars." *Sky & Telescope,* October 1984.

Turtle, A. J., et al., "A Prompt Radio Burst from Super-nova 1987A in the Large Magellanic Cloud." *Nature,* May 7, 1987.

Walborn, Nolan R., "Structure in the Carina Nebula and Eta Carinae." *Sky & Telescope,* July 1977.

Walborn, Nolan R., and Theodore R. Gull, "Eta Carinae's Numbered Days." *Sky & Telescope,* July 1982.

Wali, Kameshwar C., "Chandrasekhar vs. Eddington—an Unanticipated Confrontation." *Physics Today,* October 1982.

Welther, Barbara, "Annie Jump Cannon: Classifier of the Stars." *Mercury,* January/February 1984.

Wheeler, J. Craig, and Ken'Ichi Nomoto, "How Stars Explode." *American Scientist,* May/June 1985.

Will, Clifford, "The Binary Pulsar: Gravity Waves Exist." *Mercury,* November/December 1987.

Woosley, Stan E., and M. M. Phillips, "Supernova 1987A!" *Science,* May 6, 1988.

Other Publications

Kellermann, K., and B. Sheets, eds., "Serendipitous Discoveries in Radio Astronomy." Workshop No. 7. National Radio Astronomy Observatory, May 1983.

Philip, A. G. Davis, and David H. DeVorkin, eds., "In Memory of Henry Norris Russell." Dudley Observatory Reports No. 13. Dudley Observatory, Albany, N.Y., December 1977.

Woosley, Stan E., et al., "SN 1987a: Her True Story." Press release. Lick Observatory, University of California, Santa Cruz, June 17, 1987.

INDEX

ACKNOWLEDGMENTS

The editors wish to thank Marcia Bartusiak, Norfolk, Va.; Robin Bates, Los Angeles; Jocelyn Bell, Royal Observatory, Edinburgh, Scotland; Hans Bethe, Cornell University, Ithaca, N.Y.; Alan N. Bunner, NASA Headquarters, Washington, D.C.; Geoffrey Burbidge, Margaret E. Burbidge, Center for Astrophysics and Space Sciences, La Jolla, Calif.; Thomas H. Callen III, National Air and Space Museum, Washington, D.C.; Von Del Chamberlain, Hansen Planetarium, Salt Lake City; S. Chandrasekhar, Enrico Fermi Institute, Chicago; Geoffrey R. Chester, National Air and Space Museum, Washington, D.C.; Edward L. Chupp, University of New Hampshire, Durham; Brenda G. Corbin, U.S. Naval Observatory Library, Washington, D.C.; James Cornell, Harvard-Smithsonian Center for Astrophysics, Cambridge, Mass.; Arlin P. S. Crotts, University of Texas, Austin; Edward E. Fenimore, Los Alamos Scientific Lab, Los Alamos, N. Mex.; William A. Fowler, California Institute of Technology, Pasadena; Gordon P. Garmire, Pennsylvania State University, University Park; Martha L. Hazen, Harvard College Observatory, Cambridge, Mass.; Peter Hingley, Royal Astronomical Society, London; Fred Hoyle, Cumberland, England; Robert P. Kirshner, Harvard-Smithsonian Center for Astrophysics, Cambridge, Mass.; Yoji Kondo, NASA Goddard Space Flight Center, Greenbelt, Md.; David Malin, Anglo-Australian Observatory, Australia; Alfred K. Mann, University of Pennsylvania, Philadelphia; F. D. Miller, University of Michigan, Ann Arbor; Harvey Moseley, NASA Goddard Space Flight Center, Greenbelt, Md.; Claus Reppin, Max Planck Institut für Physik und Astrophysik, Garching, FRG; Julia L. R. Saba, NASA Goddard Space Flight Center, Greenbelt, Md.; Janet Sandlands, Royal Observatory, Edinburgh, Scotland; Gary Dean Schmidt, University of Arizona, Tucson; Ronald A. Schorn, Sky Publishing, Cambridge, Mass.; Martin Schwarzschild, Princeton University Observatory, Princeton, N.J.; Frederick Seward, Harvard-Smithsonian Astrophysical Observatory, Cambridge, Mass.; Gregory Shelton, U.S. Naval Observatory Library, Washington, D.C.; George Sonneborn, NASA Goddard Space Flight Center, Greenbelt, Md.; Edward C. Stone, California Institute of Technology, Pasadena; Jean H. Swank, NASA Goddard Space Flight Center, Greenbelt, Md.; A. J. Turtle, University of Sydney, New South Wales, Australia; Richard M. West, European Southern Observatory, Garching, FRG; Cathy White, "Nova," WGBH TV, Boston; Fred C. Witteborn, NASA Ames Research Center, Moffett Field, Calif.

PICTURE CREDITS

The sources for the illustrations that appear in this book are listed below. Credits from left to right are separated by semicolons, from top to bottom by dashes.

Cover: Courtesy Royal Observatory, Edinburgh. Front and back endpapers: Art by John Drummond. 1: Larry Sherer, from *Celestial Atlas,* by Alexander Jamieson, 1822, courtesy U.S. Naval Observatory. 2-4: Courtesy Royal Observatory, Edinburgh, inset Larry Sherer, from *Celestial Atlas,* by Alexander Jamieson, 1822, courtesy U.S. Naval Observatory. 5, 6: David F. Malin/Anglo-Australian Observatory, inset Larry Sherer, from *Celestial Atlas,* by Alexander Jamieson, 1822, courtesy U.S. Naval Observatory. 7: Courtesy Royal Observatory, Edinburgh, inset Larry Sherer, from *Celestial Atlas,* by Alexander Jamieson, 1822, courtesy U.S. Naval Observatory. 8: Larry Sherer, from *Celestial Atlas,* Alexander Jamieson, 1822, courtesy U.S. Naval Observatory. 14, 15: F. D. Miller, Department of Astronomy, University of Michigan (Ann Arbor). 16: Computer-generated initial cap by John Drummond, detail from photo appearing on pages 14, 15. 18, 19: Special effects photography by Ken Novak and Richard Berry, courtesy *Astronomy* magazine. 22, 23: Art by Mark Robinson. 25-29: Harvard College Observatory. 31: Art by Nick Schrenk. 34-37: Art by Damon M. Hertig and Daniel Rodriguez, The Art Connection. 38, 39: Dennis di Cicco, of *Sky and Telescope*—art by Damon M. Hertig and Daniel Rodriguez, The Art Connection. 40, 41: David F. Malin/Anglo-Australian Observatory. 42: Computer-generated initial cap by John Drummond, detail from photo appearing on pages 40, 41. 43: Chip Clark, courtesy Museum of Natural History, Smithsonian Institution. 45: AIP Niels Bohr Library. 46: Katherine Haramundanis. 48, 49: Courtesy Cornell University. 51-53: Art by Damon M. Hertig and Daniel Rodriguez, The Art Connection. 57: Courtesy S. Chandrasekhar. 61-69: Art by David Jonason, The Pushpin Group. 70, 71: Courtesy Royal Observatory, Edinburgh. 72: Computer-generated initial cap by John Drummond, detail from photo appearing on pages 70, 71. 74: Courtesy Royal Astronomical Society, London. 75: Courtesy Von Del Chamberlain. 76: Courtesy U.S. Naval Observatory—Orren Jack Turner, courtesy AIP Niels Bohr Library. 79-80: Courtesy the Archives, California Institute of Technology, except photo of Fred Hoyle, courtesy AIP Niels Bohr Library, E. E. Salpeter Collection. 82, 83: Courtesy *Astronomy* magazine; European Southern Observatory, Garching, FRG, inset Arlin Crotts, McDonald Observatory—Molonglo Observatory Synthesis Telescope/University of Sydney. Decorative details by Mark Robinson. 84, 85: Artwork and decorative detail by Mark Robinson, except lower left, courtesy Caltech Space Radiation Laboratory, and upper right, courtesy NASA. 87: Roger Ressmeyer/Starlight. 89: David F. Malin/Anglo-Australian Observatory. 90-101: Art by Damon M. Hertig and Daniel Rodriguez, The Art Connection. 102, 103: Lick Observatory, University of California, Santa Cruz. 104: Computer-generated initial cap by John Drummond, detail from photo appearing on pages 102, 103. 105: National Optical Astronomy Observatories. 107: Courtesy Dr. Jocelyn Bell Burnell. 111-115: Art by Stephen R. Wagner. 117: Homer Sykes/Woodfin Camp and Associates. 118: Dennis di Cicco of *Sky and Telescope,* inset Al Freni for *Life.* 120, 121: Cerro Tololo Inter-American Observatory—National Optical Astronomy Observatories and Jet Propulsion Laboratory; Dr. G. Herbig, University of California, appeared in "Symbiotic Stars," by M. Kafatos and A. G. Michalitsianos, *Scientific American,* July 1984, Vol. 251; Dr. Bradford A. Smith, University of Arizona and Dr. Richard J. Terrile, Jet Propulsion Laboratory; courtesy NRAO/AUI. 124-135: Art by Stephen Bauer of Bill Burrows and Associates.

Time-Life Books Inc.
is a wholly owned subsidiary of
TIME INCORPORATED

FOUNDER: Henry R. Luce 1898-1967

Editor-in-Chief: Jason McManus
Chairman and Chief Executive Officer:
J. Richard Munro
President and Chief Operating Officer:
N. J. Nicholas, Jr.
Editorial Director: Ray Cave
Executive Vice President, Books: Kelso F. Sutton
Vice President, Books: George Artandi

TIME-LIFE BOOKS INC.
EDITOR: George Constable
Executive Editor: Ellen Phillips
Director of Design: Louis Klein
Director of Editorial Resources: Phyllis K. Wise
Editorial Board: Russell B. Adams, Jr., Dale M.
Brown, Roberta Conlan, Thomas H. Flaherty, Lee
Hassig, Donia Ann Steele, Rosalind Stubenberg
Director of Photography and Research:
John Conrad Weiser
Assistant Director of Editorial Resources:
Elise Ritter Gibson

PRESIDENT: Christopher T. Linen
Chief Operating Officer: John M. Fahey, Jr.
Senior Vice Presidents: Robert M. DeSena, James
L. Mercer, Paul R. Stewart
Vice Presidents: Stephen L. Bair, Ralph J. Cuomo,
Neal Goff, Stephen L. Goldstein, Juanita T.
James, Hallett Johnson III, Carol Kaplan, Susan
J. Maruyama, Robert H. Smith, Joseph J. Ward
Director of Production Services: Robert J.
Passantino

Editorial Operations
Copy Chief: Diane Ullius
Production: Celia Beattie
Library: Louise D. Forstall

Correspondents: Elisabeth Kraemer-Singh (Bonn);
Maria Vincenza Aloisi (Paris); Ann Natanson
(Rome). Valuable assistance was also provided
by Judy Aspinall, Christine Hinze (London); John
Dunn (Melbourne); Christina Lieberman (New
York).

VOYAGE THROUGH THE UNIVERSE

SERIES DIRECTOR: Roberta Conlan
Series Administrator: Judith W. Shanks

Editorial Staff for *Stars*
Designer: Ellen Robling
Associate Editor: Blaine Marshall (pictures)
Text Editor: Pat Daniels
Researchers: Patti H. Cass, Karin Kinney, Edward
O. Marshall
Writer: Esther Ferington
Assistant Designer: Barbara M. Sheppard
Editorial Assistant: Alice T. Marcellus
Copy Coordinator: Darcie Conner Johnston
Picture Coordinator: Richard Karno

Special Contributors: Ronald H. Bailey, Ken
Croswell, James L. Dawson, Ann Gibbons, Peter
Gwynne, Gina Maranto, John I. Merritt, Chuck
Smith, Gerrit Verschuur, Mitchell M. Waldrop
(text); Vilasini Balakrishnan, Andrea Corell, Diana
Davis, Sanjoy Ghosh, A. R. Hogan, Jane Jacobs,
Jocelyn Lindsay, James Lomuscio, Thomas
Sodroski, Cindy Spitzer (research); Barbara L.
Klein (index).

CONSULTANTS
DAVID DeVORKIN is Curator, History of Astronomy, at the National Air and Space Museum, Smithsonian Institution. His interests include the origins
of modern astrophysics.

ELI DWEK is an astrophysicist in the Infrared Astrophysics Branch at NASA Goddard, where he
studies stellar evolution.

ICKO IBEN, JR., a professor of astronomy and physics at the University of Illinois, is interested in stellar evolution.

JAMES B. KALER, principal consultant for the volume, has taught at the University of Illinois in the
field of stellar astronomy for more than twenty
years.

MINAS KAFATOS is a theoretical astrophysicist at
George Mason University whose research interests
include black holes.

STEPHEN MARAN, a senior staff scientist at NASA
Goddard Space Flight Center, writes widely in astronomy and is press officer for the American Astronomical Society.

JEFFREY McCLINTOCK is an astrophysicist at the
Center for Astrophysics, where his investigations
cover x-ray sources and black holes.

STEN ODENWALD is an infrared astronomer at the
Naval Research Laboratory in Washington, D.C.

WILLIAM PARKE is a physics professor at the
George Washington University, where he teaches
introductory astronomy.

J. CRAIG WHEELER is chairman of the Department
of Astronomy at the University of Texas at Austin.
Much of his research is on supernovae.

STANFORD WOOSLEY is professor of astronomy
and astrophysics at the University of California at
Santa Cruz. His chief research interests include theoretical models for supernovae and the origin of
elements in stars.

**Library of Congress Cataloging in
Publication Data**
Stars/by the editors of Time-Life Books.
p. cm. (Voyage through the universe).
Bibliography: p.
Includes index.
ISBN 0-8094-6858-1.
ISBN 0-8094-6859-X (lib. bdg.).
1. Stars. I. Time-Life Books. II. Series.
QB801.S724 1989
523.8—dc19   88-20096 CIP

For information on and a full description of any
of the Time-Life Books series, please call 1-800-
621-7026 or write:
Reader Information
Time-Life Customer Service
P.O. Box C-32068
Richmond, Virginia 23261-2068

Earth: diameter 7,926 miles

Neptune: diameter 30,700 miles

Uranus: diameter 31,600 miles

Red supergiant: diameter 400 million miles

Solar System: diameter 7.5 billion miles

Globular cluster: diameter 2×10^{14} miles

Milky Way: diameter 100,000 light-years

Local Group of galaxies: 6 million light-years across

Largest double radio source: length 17 million light-years